THE
FLAVORFUL
KITCHEN
COOKBOOK

© 2013 Fair Winds Press
Text © 2010 Robert Krause and Molly Krause
Photography © 2010 Rockport Publishers
This edition published in 2013

First published in the USA in 2013 by
Fair Winds Press, a member of
Quayside Publishing Group
100 Cummings Center
Suite 406-L
Beverly, MA 01915-6101
www.fairwindspress.com

17 16 15 14 13   1 2 3 4 5

ISBN: 978-1-59233-589-3
Digital edition published in 2013
eISBN: 978-1-61058-866-9
This book was originally published with the title
The Cook's Book of Intense Flavors (Fair Winds Press, 2010).

Originally found under the following Library of Congress Cataloging-in-Publication Data
Molly, Krause.
  The cook's book of intense flavors : 101 surprising flavor
combinations and extraordinary recipes that excite your palate and
pleasure your senses / Molly and Robert Krause.
      p. cm.
  Includes index.
  ISBN-13: 978-1-59233-432-2
  ISBN-10: 1-59233-432-6
  1. Cooking, International. I. Krause, Robert. II. Title.

TX725.A1M64 2010
641.59--dc22

                          2010020270

Cover design by Quayside Publishing Group
Book design by Kathie Alexander
Photography by Luciana Pampalone

Printed and bound in China

THE FLAVORFUL KITCHEN COOKBOOK

101 AMAZING 3-INGREDIENT FLAVOR COMBINATIONS

ROBERT and MOLLY KRAUSE, chef/owners of the award-winning Krause Dining restaurant, featured in *Food & Wine* magazine's annual "100 Tastes to Try" issue

FAIR WINDS
PRESS
BEVERLY, MASSACHUSETTS

CONTENTS

INTRODUCTION

8   Welcome to the World of Flavor

CHAPTER 1 Timeless with a Twist

Combinations that make the familiar new again.

14   Bacon + Egg + Mango

16   Spinach + Cheese + Anise Liqueur

18   Apple + Horseradish + Vanilla

21   Asparagus + Nut + Cheese

24   Zucchini + Nasturtium + Dill

26   Fig + Cantaloupe + Passion Fruit

28   Green Curry + Vanilla + Cucumber

31   Tomato + Cheese + Truffle

34   Parsnip + Vanilla + Lavender

36   Chile Pepper + Coconut + Tomato

38   Peppercorn + Coriander + Cardamom

41   Chocolate + Ginger + Nut

44   Soda Pop + Liquor + Cream

46   Caramel + Corn + Chile Pepper

CHAPTER 2 Unexpected Pleasures

*Combinations that may seem unusual,
but make sense in an innovative new way*

51   Cumin + Cinnamon + Cocoa

54   Fig + Apple + Anchovy

56   Tomato + Brown Sugar + Coffee

58   Watermelon + Cheese + Vinegar

61   Cocoa + Red Wine + Mint

64   Chestnut + Miso + Orange

66   Mushroom + Rose + Lavender

69   Corn + Chile Pepper + Rosemary

72   Goat's Milk + Rosemary + Rhubarb

74   Chocolate + Bacon + Brown Sugar

77   Chocolate + Vanilla + Chile Pepper

CHAPTER 3 Complex Concoctions
Combinations with deep and distinctive flavors.

82	Fig + Cheese + Bacon
84	Sunchoke + Cheese + Bitter Greens
86	Parsnip + Ginger + Basil
89	Lobster + Cream + Smoked Paprika
92	Poppy Seed + Sesame Seed + Onion Seed
94	Salt + Fennel + Melon
96	Egg + Chile Pepper + Prosciutto
99	Tomato + Fennel + Mustard
102	Apple + Honey + Almond
104	Onion + Carrot + Celery
106	Blue Cheese + Pear + Nut
109	Lemon + Fennel + Nut
112	Cherry + Yogurt + Sesame

CHAPTER 4 Bright & Light
Combinations that lend a delicate and fresh flavor to the final dish.

116	Apple + Fennel + Lemon
118	Cucumber + Beet + Cabbage
121	Berry + Citrus + Cheese
124	Watermelon + Tomato + Mint
126	Cucumber + Basil + Mint
128	Orange + Saffron + Vanilla
131	Asparagus + Orange + Oregano
134	Carrot + Fennel + Wine
136	Grapefruit + Onion + Parsley
139	Avocado + Mango + Nut
142	Parsley + Garlic + Lemon
144	Pear + Cucumber + Leek
147	Grape + Saffron + Nut
150	Apricot + Honey + Thyme
152	Rhubarb + Ginger + Lemon

CHAPTER 5 Sweet & Sour

*Combinations that pursue the sweet-and-sour
element pervasive in many world cuisines.*

156   Cauliflower + Turmeric + Vinegar

158   Peach + Ginger + Vinegar

161   Cucumber + Jalapeño + Vinegar

164   Apricot + Brandy + Cardamom

166   Radish + Peppercorn + Rose Hip

168   Raisin + Mustard Seed + Vinegar

171   Mango + Cinnamon + Vinegar

174   Jicama + Cabbage + Sesame

176   Cabbage + Dill + Cream

178   Bean + Bacon + Vinegar

180   Currant + Chile Pepper + Vinegar

183   Tea + Lime + Sugar

186   Coffee + Fig + Vinegar

CHAPTER 6 Exotic Flavors

Combinations inspired by ethnic world cuisines.

190   Peanut + Soy Sauce + Chile Pepper

192   Turnip + Miso + Mirin

195   Lemongrass + Coconut + Basil

198   Cilantro + Cumin + Turmeric

200   Plantain + Curry + Vanilla

203   Date + Chestnut + Paprika

206   Orange + Cardamom + Coconut

208   Chickpea + Edamame + Coriander

211   Sumac + Sesame + Thyme

214   Pumpkin Seed + Chile Pepper + Garlic

216   Cactus + Chayote + Apple

218   Avocado + Jicama + Chile Pepper

220   Bottarga + Garlic + Lemon

223   Coffee + Cardamom + Pistachio

226   Olive Oil + Citrus + Wine

228   Okra + Fennel + Coriander

230   Squash + Red Curry + Nut

232   Curry + Apple + Coconut

CHAPTER 7 Decidedly Decadent

Combinations that seek the biggest, richest flavor possible.

236   Truffle + Pepper + Butter

238   Liver + Pea + Sage

241   Bacon + Apple + Cognac

244   Quince + Mango + Chile Pepper

246   Egg + Caviar + Chervil

248   Chocolate + Lime + Cream

250   Pumpkin + Foie Gras + Truffle

253   Salt + Pepper + Sugar

255   Chocolate + Chocolate + Cream

258   Marrow + Citrus + Parsley

260   Mushroom + Liver + Cream

262   Banana + Chocolate + Hazelnut

264   Pear + Chocolate + Hazelnut

267   Lemon + Raspberry + Egg

270   Chocolate + Cheese + Berry

272   Egg + Liquor + Nutmeg

274   Chocolate + Peanut Butter + Malt

277   **CONCLUSION**

277   **RESOURCES**

279   **ACKNOWLEDGMENTS**

281   **ABOUT THE AUTHORS**

282   **RECIPE INDEX**

285   **GENERAL INDEX**

INTRODUCTION

Welcome to the World of Flavor

Why should you bother with another cookbook? You know your way around a kitchen, feed you and your loved ones home-cooked meals regularly, and even have some impressive dishes under your belt to prepare for company. You recognize most of the ingredients at the market, even the gourmet ones, and may even subscribe to a cooking magazine or two. You already know how to cook.

And yet.

The act of cooking is really like engaging in a relationship, except that instead of another person, food becomes the object of your attention. And just like people, food can be complicated. Some personalities just don't get along (such as olive oil with Asian cuisine); sometimes you can be pleasantly surprised despite a negative first impression (turnips can be delicious); or sometimes, timing is everything (as in adding fresh herbs at the end of a cooking process). Just as we should make it a priority to strive to improve our relationships with those close to us, we can always deepen our relationship with the food we prepare.

This book approaches dishes in a different way than most cookbooks. Most cookbooks break down by type of recipe and when during the meal it appears (for example, salads, soups, main courses, and desserts), but this book looks at recipes through the lens of flavor combinations. Each recipe highlights three flavors featured in the dish, with individual discussions of each, including cooking techniques that work well and seasonality, if applicable, as well as the combination of flavors as a whole in terms of how they interact with each other. We also offer additional ways to use the flavors together beyond the recipe provided. If these three flavors like to hang out together, we want to tell you all that they can do!

We didn't group the recipes in this book by dish type—there is no "appetizer" chapter. Rather, the chapters try to cluster flavor profiles to evoke a sense of taste that can often be hard to describe. Take, for example, Chapter 2, "Unexpected Pleasures." These combinations have a surprising element or component that sounds as though it may not go with the others. The recipes range from spice blends to sauces and even desserts. Each chapter has a theme to unite the flavor combinations but will otherwise run wild with a wide range of recipes that will hopefully draw you back into the kitchen to pursue.

Cooking is about combining flavors as well as applying different preparation techniques, and that is what this book is all about. Let your mood guide where to jump in—the book has no real beginning or end. Are you feeling restless but have nowhere to go? Check out Chapter 6, "Exotic Flavors." Do you have a few days off from work and want to tackle a new and challenging recipe? Look into Chapter 3, "Complex Creations." Are you feeling a craving for something rich? How about Chapter 7, "Decidedly Decadent"? You get the idea.

In addition to your mood, let the food choose you for a dish. Some beautiful mushrooms may call to you from their baskets, or a pile of heirloom tomatoes may look irresistible. Give in! Then let this book help you figure out what to do with them. Seek to make your cooking routine more fun and spontaneous, while keeping in place solid cooking techniques.

Intense cooking contains rich flavors, and we go for maximum flavor in this collection of recipes. You may find some ingredients repeated throughout the chapters. That is because some ingredients, such as chile peppers for example, have great potential for intensity and diversity. Other ingredients, such as rhubarb, may not have as diverse applications, but they offer a distinct flavor worthy of appreciation.

So go for flavor in your cooking and bring with it your own intensity—and expect great combinations to result.

CHAPTER 1

Timeless with a Twist

Certain flavor combinations are so ingrained in our collective consciousness, it seems that they have always existed together—salt with pepper, bacon with eggs, peanut butter with jelly. We take them for granted and almost regard them as indistinguishable from their individual states. Part of their charm lies in their mass appeal and in not having to wonder whether someone will enjoy these flavors, because overall everyone does.

And yet, just as a timeless piece of clothing sometimes needs a new accessory to keep it relevant, timeless food combinations also benefit from a breath of fresh air. We seek to keep the recognizable quality in these flavors, but we also want to bring in a twist to keep the tasting interesting.

Timeless food combinations typically fall into one of two categories: The ingredients themselves are frequently prepared together, or the method of preparation (or cooking technique) sets the standard for serving. To change up these classics, we either introduce an ingredient not normally found in the preparation, or we prepare the standard ingredients in a different way using an untraditional technique.

Consider the classic combination of bacon and eggs. We appreciate the salty meatiness of bacon with many varieties of cooked eggs to provide a substantial and satisfying breakfast. Add tropical mango, and now we experience refreshment—a "twist" in the form of Apple-Smoked Bacon Frittata with Mango Chutney (page 15).

Or how about the summer staple basil pesto that flavors many pasta and vegetable dishes and uses up our garden basil surplus? For an alternative, we prepare a pesto made from asparagus ends, reserving the highly valued tips for another purpose. This part of asparagus, a portion typically not valued highly, gives a great flavor when combined with the recognizable pesto additions of nuts and cheese. It's classic pesto—with a twist!

Finally, consider the tomato, whose preparations seem almost limitless—raw, sautéed, puréed, grilled, fried, and more. Combining them with chile pepper is nothing new, but introducing coconut for a sweet element in Chile-Spiked Lentils (page 37) offers a different take on the typical tomato preparations.

So forge ahead with familiar flavors and combine them in different ways using altered cooking techniques for a taste of the timeless—with a twist.

"BRUNCH IN BALI"

BACON + EGG + MANGO

The classic American pairing of bacon and eggs satisfies countless breakfast eaters. Eggs' versatility makes them easy to use, and bacon's salty protein provides a welcome addition. Here, the surprise of the fruity mango both lightens and intensifies the familiar combination. Because eggs adapt so well, let their transformation lead the way into different applications. This recipe bakes the eggs into a frittata; also try hard-boiled eggs with bacon and mango over greens for a fresh take on a Cobb salad.

BACON

For this combination, seek out bacon with some additional flavor components to play off of the mango. Apple-smoked bacon or maple-roasted bacon both offer intensity that could work well.

EGG

An egg can literally transform a dish (think soufflé), use its yolks only (think aioli), or whites only (think meringue). And of course, it can be baked, poached, fried, scrambled, and used extensively for baking.

MANGO

Fresh, ripe mango is luscious unadorned in fruit salads or puréed into smoothies and chilled soups. Cooking heightens the sugars and works nicely in compotes for fish or desserts. Dried mango pairs well with granola and for snacking. In this combination, think of it as an accent piece, an unexpected bit of sultry fruit.

"BRUNCH IN BALI" RECIPE

Apple-Smoked Bacon Frittata with Mango Chutney

Baked egg dishes such as frittatas are among the easiest to prepare and are especially useful for feeding a group. Frittatas are also easily improvised, so experiment with whatever cheese or vegetables you have in your refrigerator.

Mango Chutney

4 **mangoes**, peeled and diced

Juice from 1 lemon

1 cup (235 ml) dry fruity white wine
   (such as Riesling)

¾ cup (150 g) sugar

¼ vanilla bean, split lengthwise
   and seeds scraped

Salt

Frittata

8 ounces (227 g) **apple-smoked bacon**

1 small yellow onion, diced

6 cups (180 g) baby spinach

12 **eggs**

1 teaspoon (6 g) salt

1 cup (100 g) grated Parmesan
   cheese, divided

🍴 A Sunday brunch will never be the same once you pull out dynamite dishes such as this one. When you have beautifully ripe mango, make some extra chutney to keep around for toast in the morning.

To prepare chutney: Divide the diced mango and put into half into a saucepan. Set aside the other half. Add the wine, sugar, vanilla bean, and a pinch of salt to the saucepan and cook until the liquid reduces by half and the mango softens.

Purée cooked mango and liquid in a food processor until smooth. Return smooth purée to saucepan and add remaining uncooked mango. Cook over low heat until remaining fruit becomes soft and chutney thickens slightly. Allow to cool, cover, and keep refrigerated until ready to use. (Chutney may be refrigerated for up to 1 week.)

To prepare frittata: Preheat oven to 350°F (180°C, or gas mark 4). Cook bacon in a 12-inch (30 cm) nonstick ovenproof skillet over medium heat until crisp. Remove bacon with a slotted spoon and drain on paper towel–lined plate. Remove 1 tablespoon (15 ml) bacon fat from the skillet and set aside. Using the bacon fat remaining in the skillet, cook onion until translucent. Add spinach and cook until wilted, 1 minute. Remove spinach from skillet and allow to cool on a plate. Wipe the skillet with a paper towel to remove the remaining juices.

In a large bowl, beat the eggs and salt. Whisk in half of the bacon, half of the Parmesan, and all of the spinach. Heat the set aside bacon fat in the skillet and pour in the egg mixture. Sprinkle the remaining bacon and cheese over the top. Cook until the edges set. Transfer to the oven and bake until completely set. Remove and transfer to a serving platter. Allow to cool for 20 minutes. Slice and serve with mango chutney.

Yield: 6 servings, plus 2 cups (500 g) chutney

"FRESH AND ZINGY"

SPINACH + CHEESE + ANISE LIQUEUR

Spinach offers a clean, strong vegetal flavor of which everyone knows the health benefits. Despite that, few of us consume it frequently or in large quantities. When cooked, the acid in spinach increases, so spinach is often paired with a rich ingredient to offer balance. In this combination, cheese is used for that richness. Anise liqueur also naturally accents the spinach.

The application recipe prepares a version of a Rockefeller topping, but the combination also works well as a filling in pasta dishes (such as ravioli), baked into lasagna, or tossed with penne. This ingredient trio tastes particularly good with seafood such as oysters, scallops, and salmon, as well as with chicken or lamb.

SPINACH

Look for it fresh year-round, but don't overlook its frozen form for long storage or use in casseroles and dips. Remember when cooking raw spinach that it reduces in size dramatically once cooked. Also, because fresh spinach holds moisture (released when cooked), you may need to drain or squeeze it before mixing with other ingredients; otherwise, your dish may end up unpleasantly watery.

CHEESE

Let your cooking technique drive your selection of cheese. Emmenthaler makes a nice choice in baking for a bubbly, melted result. Ricotta is great filling for pasta and lasagna. Grated Parmigiano-Reggiano or Pecorino Romano always are flavorful additions, especially broiled with this combination.

ANISE LIQUEUR

The licorice taste of anise liqueur is a winning match with spinach. The French-made Pernod or Herbsaint, originally made in New Orleans, are both great choices for this combination. Also, the Greek alcohol ouzo has anise flavor, so it's another possible option. Be sure to add the liqueur at the end, after cooking, to maximize flavor effect.

"FRESH AND ZINGY" RECIPE

Beyond Oysters Rockefeller

This topping is so flavorful and versatile that it is a shame not to use it for more dishes than oysters. See variations at the end of the recipe for specific applications.

3 slices pancetta, diced small

1 large shallot, diced

2 cloves garlic, diced

2 tablespoons (28 g) unsalted butter

1 package (12 ounces, or 340 g) **frozen chopped spinach**, thawed, liquid squeezed (or substitute 1 pound [455 g] **fresh chopped spinach**, blanched and liquid squeezed)

¼ cup (60 ml) dry white wine (such as Chardonnay)

2 tablespoons (30 ml) Worcestershire sauce

1 tablespoon (15 ml) red wine vinegar

½ cup (120 ml) heavy cream

1 cup (100 g) grated **Parmigiano-Reggiano**

¼ cup (60 ml) **anise liqueur** (such as Pernod)

Salt and pepper

In a large skillet, cook the pancetta over medium heat until the fat is rendered and the pancetta is just barely crisp. Pour off half the fat and then add the shallot, garlic, and butter. Reduce heat to low and cook for 5 minutes, stirring occasionally.

Add the spinach, wine, Worcestershire sauce, vinegar, and cream and cook for an additional 10 minutes over low heat. Remove from heat and stir in the cheese and liqueur. Season with salt and pepper. Prepare up to 1 day in advance. Keep refrigerated.

Yield: 2 cups (230 g)

Here are some options for using the spread: Top crostini with a dollop and a sprinkling of shaved Parmesan to serve as an appetizer; top seared scallops, shrimp, or salmon with it and broil for a minute or two; or go classic and top raw oysters on the half shell and then bake for 5 minutes.

"A KISS ON THE CHEEK AND A SLAP TO THE FACE"

APPLE + HORSERADISH + VANILLA

Sweetness, pungency, and aroma dominate this dynamic flavor combination. The heat of horseradish livens up wholesome apples. Add to that the vanilla and out comes a surprisingly pleasing savory combination. Adjust your handling of the apple for more variation: Slice and sauté it with the components to top a pot roast, grate it raw for an easy no-cook compote, or cook it together for a cream sauce, as in the application recipe. For the fastest route, use a natural store-bought applesauce and a prepared horseradish and stir in vanilla seeds.

APPLE

Your preparation of the apple determines the texture of the combination because it is the only ingredient that that can remain raw with a crisp texture. To add a crunchy element to your sauce, reserve some diced raw apple to add to the finished product. To layer the apple-flavor component, use apple juice, cider, or vinegar in your combination.

HORSERADISH

Prepared horseradish combines the root with vinegar and sometimes cream. Fresh horseradish is more difficult to find, and its intensity can vary in strength but can offer a livelier flavor. If using the root, peel and grate it using a micrograter (or a cheese grater). Refrain from putting too much into the dish. It should be an accent or condiment only.

VANILLA

Do not substitute vanilla extract for vanilla beans in this combination. In addition to the complementary aromatic qualities it offers, vanilla's visual aspect—its little black specks—is important. They add a level of intrigue and surprise in a savory dish and reveal the use of high-quality ingredients.

"A KISS ON THE CHEEK AND A SLAP TO THE FACE" RECIPE

Creamy Applesauce with Sass

This is a great sauce for lighter meats such as pork, veal, or chicken. Or try it as the flavor base to add to potato, parsnip, or sweet potato purée.

1 tablespoon (15 ml) almond oil (optional)

1 tablespoon (14 g) unsalted butter

4 **Granny Smith apples**, peeled, cored, and diced

1 small white onion, diced

1 jar (5 ounces, or 142 g) prepared **horseradish**

1 **vanilla bean**, split lengthwise and seeds scraped

¼ cup (60 ml) semisweet white wine (such as Riesling)

1 tablespoon (20 g) honey

1 sprig fresh thyme

2 bay leaves

2 cups (475 ml) heavy cream

Salt and pepper

In a large skillet, heat almond oil (if using) with butter over medium heat. Add apples and onion and cook until both turn translucent. Add horseradish and vanilla seeds and pod and stir to incorporate.

Deglaze the pan with the wine and simmer until most of the liquid evaporates. Add honey, thyme, bay leaves, and cream. Cook for 5 to 10 minutes until liquid reduces by half. Remove thyme, bay leaves, and vanilla pod and allow to cool slightly. Season with salt and pepper.

Mix in a blender until smooth. Serve immediately (rewarm if necessary) or cool and keep covered in the refrigerator for several days.

Yield: 1½ cups (355 ml)

Apples are at their peak during the fall months, a nice time to try this sauce with roasted pork tenderloin.

"SUNNY SPRING DAY"

ASPARAGUS + NUT + CHEESE

Fresh asparagus, one of the early vegetables to be harvested, is Mother Nature's signal that spring is in full swing. Substantial tastes such as nuts and cheese enhance the distinct, fresh flavor of asparagus.

This application has a balanced composition, resulting in a thick spread that plays on the traditional basil pesto. Increase the asparagus ratio to create a soup with a sprinkling of Parmesan cheese and almonds. These ingredients could also make a lovely salad over fresh greens.

ASPARAGUS
Green is the most commonly available asparagus, but this vegetable also grows in purple and—the mildest flavored—white varieties. White asparagus stays this pale hue because it is denied light while grown. Asparagus is crispy when fresh, tender and delicious when cooked, and purées nicely.

NUT
Toasting and roasting nuts intensifies their flavors, something we recommend for this combination. The recipe calls for pine nuts because they are the classic pesto nut, but almonds would work nicely as well.

CHEESE
In smaller proportions in a dish, cheese can serve as a flavor accent or a creaminess enhancer. Because it melts well, in can act as the base for sauces, dips, and soups. Generally, cooking cheese affects its texture more than its flavor. To make cheese more dominant in this mix, consider making a creamy cheese fondue sprinkled with almonds and dipping in blanched asparagus.

"SUNNY SPRING DAY" RECIPE

Welcome-the-Season Pesto

This version of pesto takes advantage of the early vegetable bounty when basil is not yet abundant in the garden. Use this asparagus pesto as a spread on crostini, with hummus and pita, or to season steamed asparagus.

2 pounds (900 g) **asparagus**
1 ounce (28 g) **pine nuts**
2 tablespoons (30 ml) almond oil (or canola oil)
Salt
2 ounces (57 g) grated **Parmesan cheese**
2 tablespoons (30 ml) grapeseed oil (or canola oil)
Pepper

Cut off and discard the woody ends of the asparagus. With the remaining asparagus, cut off the bottom 2 inches (5 cm) of usable ends. These are used for the pesto. Reserve the larger portion of asparagus for a different use.

In a skillet over medium heat, toast the pine nuts with almond oil and a pinch of salt until browned. Add asparagus, pine nuts, cheese, and grapeseed or canola oil to a food processor and pulse until rough-chopped but a spreadable consistency.

Adjust seasoning with salt and pepper. Serve or cover and refrigerate for up to 3 days.

Yield: 1 cup (260 g)

This unique application uses only the end sections of the asparagus. Layer your flavors by serving the pesto with the blanched asparagus tips for a simple preparation or by making soup with the remaining asparagus.

"DYNAMIC GARDEN FLAVOR"

ZUCCHINI + NASTURTIUM + DILL

This garden harvest of ingredients comes together to create fresh, vibrant flavors. The peppery nasturtium and the crisp, grassy dill weed enhance the sweet zucchini.

The application recipe prepares crispy zucchini-dill fritters with a creamy nasturtium aioli. Here's another option for this trio: Sauté zucchini with dill and top it with a nasturtium butter. Or stuff zucchini and nasturtium blossoms with a dill-yogurt filling. These flavors taste great with lemon, orange, creamy cheeses and yogurt, and other vegetables, such as summer squash.

ZUCCHINI

It is not necessary to peel zucchini before grating them (especially those grown without chemicals), but be sure to wash them. For large ones, split in half lengthwise and scoop out the seeds; the smaller ones have seeds too small to notice, so use the whole vegetable. The easiest way to grate is to use a food processor—chop into large pieces and pulse until grated. A box grater does the job as well. Be sure to drain in a colander or with paper towels, as zucchini release water when grating. Grated zucchini freezes well in sealed plastic bags.

NASTURIUM

Cooking applications use both the leaves and blossoms of nasturtium. Its leaves are similar to watercress in that they are spicy; the blossoms are a bit like radish but sweeter and tangier. Using nasturtium raw or lightly heated results in the most vibrant, zippy flavor.

DILL

Dill weed is prized for its clean, pleasant flavor when fresh (don't bother with the too-mild dried dill). Dill weed (the frilly green ends) and dill seed (the small dried balls) differ significantly in flavor, so note which a recipe suggests. Fresh dill weed should not be cooked for an extended time—exposure to heat diminishes its flavor. But toasting accentuates dill seed's flavor.

"DYNAMIC GARDEN FLAVOR" RECIPE

Crispy Zucchini with Peppery Aioli

You can make the aioli up to a couple days ahead of serving, and use any leftovers on steamed fish or vegetables.

Aioli

2 cloves garlic, peeled

Zest from 1 orange

2 egg yolks

1 teaspoon (4 g) Dijon mustard

⅛ teaspoon Sriracha or hot sauce
(optional)

1 cup (235 ml) peanut oil

Juice from ¼ lemon

1 cup (116 g) packed **nasturtium
blossoms**, washed and dried

2 tablespoons (30 g) plain yogurt,
preferably Greek style

Salt and pepper

Fritters

1 pound **zucchini** (455 g), grated
and drained

1 red onion, diced

2 tablespoons (8 g) chopped fresh
dill weed

1 tablespoon (2 g) chopped
nasturtium leaves

½ cup (60 g) all-purpose flour

2 eggs, lightly beaten

Salt and pepper

½ cup (120 ml) olive oil, for frying

To prepare aioli: In a food processor, pulse the garlic and orange zest until finely chopped, stopping to scrape the sides once. Add the egg yolks, mustard, and Sriracha (if using) and blend until smooth and creamy.

With the motor running, add the peanut oil a little at time and then add the lemon juice. It should look thick and emulsified. Add the nasturtium blossoms and yogurt and process until thick and creamy. Season with salt and pepper. Use immediately or keep covered in the refrigerator for several days.

To prepare fritters: Squeeze zucchini in paper towels to remove any water. Mix with onion, dill, nasturtium, flour, and eggs and season with salt and pepper.

In a large skillet, heat olive oil over medium heat. Drop fritter batter by the heaping tablespoon and fry until golden brown on each side, 2 to 4 minutes per side. Drain on paper towels and serve hot topped with a dab of aioli.

Yield: 4 servings

This recipe prepares a small fritter to serve as an appetizer or small course. Feel free to make the fritters larger, serve with a salad, and make them a meal.

"SWEET AND SUCCULENT"
FIG + CANTALOUPE + PASSION FRUIT

This trio of fruit offers a complexity of flavors sure to produce delicious eating. The succulent, luscious fig complements the refreshingly sweet cantaloupe. The tart passion fruit offers an accent flavor that elevates the entire combination. Cook these flavors together for a jam or compote, purée them into a shake or smoothie, or make into a vinaigrette. Cured hams such as prosciutto or Serrano, as well as chicken livers (as in a pâté or mousse), make flavorful additions to the combination.

FIG

Fresh figs enjoy a short season from late summer throughout the fall months. They are also dried and are effectively rehydrated. Keep the liquid (usually water or wine) used to rehydrate dried figs for use in sauces and vinaigrettes or reduced to a syrup glaze. For fresh varieties and flavor versatility, look for the dark purple Black Mission or the light green Kadota.

CANTALOUPE

Also known as muskmelons, cantaloupes are ripe when heavy and show very little green tinge on the outer rind. When puréed, they have a smooth consistency, excellent for smoothies and chilled soups. In this combination, substitute honeydew melon or papaya for the cantaloupe, if desired.

PASSION FRUIT

Small and wrinkly, fresh passion fruits contain scores of edible seeds that often need to be strained for sauces. Their tartness usually requires addition of a sweetener and creates a wonderfully fragrant flavor. You can substitute fresh passion fruit with frozen concentrate (often found with Mexican products), but taste before adding sugar, as the concentrates often already have sugar added.

"SWEET AND SUCCULENT" RECIPE

Melon and Prosciutto with Fig Syrup and Passion Fruit Sorbet

This composed salad requires several separate steps, but you can do everything in advance. Prepare the fig syrup and the passion fruit sorbet a couple of days ahead, making it easy to plate up for guests.

Special equipment required

Fig Syrup

1 cup (235 ml) port wine

½ cup (100 g) sugar

7 fresh **figs** (preferably Black Mission), halved, or 8 dried figs, stems cut off

Passion Fruit Sorbet

½ cup (75 g) **passion fruit** purée (frozen or strained fresh pulp)

½ to 1 cup (120 to 235 ml) simple syrup (page 183); taste to determine

To Serve

1 pound (455 g) prosciutto, thinly sliced

1 **cantaloupe**, cut into 8 flat rectangular pieces, ½-inch (1¼ cm) thick

4 sprigs watercress

To make fig syrup: In a saucepan, combine port, sugar, and figs; bring to a simmer. Cook until liquid reduces to ½ cup (120 ml). Let cool briefly and then purée in a food processor or blender just to break up the figs—the finished syrup should contain visible pieces of figs. Cover and refrigerate for up to 1 week.

To make the passion fruit sorbet: Mix passion fruit and simple syrup, tasting to ensure a sweet, tart flavor. Add to an ice cream machine and twist according to manufacturer's instructions. Keep covered in the freezer for up to 1 week until ready to use.

To serve: Warm fig syrup in a saucepan if it has been refrigerated. Spread 4 slices of prosciutto on each serving plate, overlapping the pieces to cover the plate. Place 2 pieces of cantaloupe on top of the prosciutto (let your inner artist be your guide), drizzle the fig syrup over the prosciutto-cantaloupe plate, and top each melon slice with a small scoop of passion fruit sorbet. Serve immediately.

Yield: 4 servings

The figs and melon are in season during late summer, and because you can use dried figs if necessary, let the quality of the melon determine when you make this dish.

"ULTRA COOL WITH AN EDGE"

GREEN CURRY + VANILLA + CUCUMBER

The wonderful, full flavor of green curry becomes more complex and interesting by adding vanilla, typically used for sweet offerings but which gives the curry a rich and pleasing aroma. The cool cucumber is a welcome and lifting presence in the combination. In the West, we typically eat our cucumbers raw, but Eastern cuisine frequently sautés or simmers cucumbers in curries, which offers additional variations for this combination.

The application recipe is a classic French beurre blanc preparation, adding the twist of the Asian curry with vanilla. Another option is to make a green curry–vanilla marinade for fish to serve with a cucumber-tomato salad. These flavors work well with seafood, chicken, tofu, tempeh, and rice.

GREEN CURRY

Green curry has the heat of red curries but also a definite and desired sweetness. Most of the heat comes from ground green chiles and seasonings such as kaffir lime and shrimp paste. You can make your own, but you can also get equally excellent results with a store-bought paste, available at Asian markets.

VANILLA

Do not substitute vanilla extract for vanilla beans in this combination. In addition to the complementary aromatic qualities it offers, vanilla's visual aspect—its little black specks—is important. They add a level of intrigue and surprise in a savory dish and reveal the use of high-quality ingredients.

CUCUMBER

When making a cucumber broth, as in this recipe, let your finished result determine how to treat the cucumber. Contrary to the cucumber broth made for Braised Halibut with Warm Cucumber Broth (page 145), which directs you to purée the whole cucumber, peel, seeds, and all, this recipe calls for peeling the cucumber first. Here's the difference: This beurre blanc recipe desires a clear, light color for the finished sauce, as opposed to keeping the peel on, resulting in a green sauce.

"ULTRA COOL WITH AN EDGE" RECIPE

Creamy, Spicy Beurre Blanc

This rich and flavorful sauce goes perfectly with fish or scallops.

2 large **cucumbers**, peeled
and chopped

1 tablespoon (18 g) salt

½ cup (120 ml) water

¼ cup (6.4 g) mint leaves

1 teaspoon (5 ml) rice vinegar

1 tablespoon (15 ml) mirin
(cooking sake)

Juice from 1 lemon

1 tablespoon (15 ml) vegetable oil
(such as canola)

2 cloves garlic, peeled and minced

1 inch (2.5 cm) fresh ginger, peeled
and minced

1 **vanilla bean**, split lengthwise
and seeds scraped

1 cup (235 ml) semisweet white wine
(such as Spätlese Riesling)

1 tablespoon (15 g) **green curry paste**

6 tablespoons (84 g) cold unsalted
butter, cut into 6 pieces

3 tablespoons (45 g) crème fraîche
(or sour cream)

Salt and pepper

In a blender, add cucumbers, salt, water, mint, vinegar, mirin, and lemon juice. Blend until smooth, 1 to 2 minutes. Strain cucumber broth through a fine mesh strainer and reserve.

In a large skillet, sauté garlic and ginger in canola oil until softened. Add vanilla bean pod and seeds, wine, curry paste, and cucumber broth. Stir, bring to a simmer, and cook for 10 minutes, reducing liquid until only ¼ cup (60 ml) remains. Remove vanilla pod and save for future use.

Whisk in the cold butter 1 piece at a time, constantly stirring and waiting to add another piece until the prior has been fully incorporated and emulsified. Remove from heat and strain through a fine mesh strainer. Whisk in crème fraîche (or sour cream, if using) and season with salt and pepper. Serve immediately.

Yield: 1 cup (235 ml)

Take your time when adding your butter to the beurre blanc sauce or you risk "breaking" it—when it separates and is no longer emulsified. (See Grilled Portobellos with Floral Beurre Blanc, page 67, for tips on a saving a broken sauce.) Make the sauce slightly in advance of serving and keep it over a warm bowl of water.

"EVERYDAY ELEGANCE"

TOMATO + CHEESE + TRUFFLE

The versatility of both tomatoes and cheese makes this a fun combination. The garden freshness of tomato coupled with the rich, unctuous cheese is delicious enough on its own, but bring into the mix the aromatic truffle and things can get a bit heady.

Our application recipe keeps it simple with a salad of ripe tomato, fresh mozzarella, and white truffle oil. For cooked applications, consider a creamy tomato soup with toasted, truffled cheese sandwiches; an omelet with tomato, Cheddar cheese, and fresh truffles; or a tomato risotto with Parmesan and truffle butter.

TOMATO

For uncooked recipes such as salads, wait until the summer, when tomatoes are ripe and preferably home grown, or experiment with some of the many colored varieties at your local farmers' market. For soups and risottos, high-quality canned products work nicely.

CHEESE

When melting cheese, as for sandwiches, choose an Italian fontina. For salads, fresh mozzarella is lovely. Bread and fry it for a nice change. To accent a dish or in risotto or pasta, Parmigiano-Reggiano can't be beat.

TRUFFLE

Because of truffles' pungent flavor—white truffles have a stronger flavor than black—and high cost, cooks use them sparingly in food. They are available seasonally fresh or jarred in a light brine. Truffle-flavored oils and butters are also readily available. To highlight the truffle element in any dish, use truffle oil with fresh shaved truffles as a finishing garnish.

"EVERYDAY ELEGANCE" RECIPE

Simple and Elegant Tomato-Basil Salad

Use the best tomatoes you can find for this salad (large or cherry work best); its straightforward preparation demands high-quality ingredients. To make your own basil oil, blanch 1 cup (35 g) packed basil leaves in boiling water for 30 seconds then purée leaves in blender with ½ cup (120 ml) olive or canola oil.

2 large **tomatoes** (preferably heirloom), in ½-inch (1¼ cm) slices
2 large balls **fresh mozzarella cheese**, in ½-inch (1¼ cm) slices
12 fresh basil leaves
Drizzle **white truffle oil** or basil oil
1 medium fresh truffle, thinly shaved
Salt and pepper

Arrange layers of tomato, mozzarella, and basil leaves on a platter or individual plates. Drizzle lightly with truffle or basil oil, scatter with shaved truffles, and season with salt and pepper. Serve immediately.

Yield: 4 servings

🍴 To make this salad more substantial, add toasted pine nuts (as pictured), Black Forest ham, poached shrimp, or garlic-rubbed toast.

"SAVORY, FLORAL, AND FRAGRANT"

PARSNIP + VANILLA + LAVENDER

The natural sweetness of parsnips makes them compatible with ingredients normally used in desserts—in this case the lovely vanilla and floral lavender. There are many savory applications, however, with smooth purées (especially nice with fish) and creamy soups. To take this to the dessert side, make a vanilla-lavender ice cream and top with sweet and crunchy parsnip chips. Other root vegetables would work well with this combination, too, such as potatoes, carrots, or celery root. Or serve it with rich cuts of beef for a nice contrast.

PARSNIP

Although it looks like a white carrot, a parsnip has a milder, sweeter flavor than its orange cousin. Use smaller, more delicate parsnips grated in salads. Otherwise, parsnips are well suited to cooking techniques—broiling, roasting, and braising to soften and increase the sugars. They also purée nicely. They are sometimes covered in wax at the grocery store—be sure to remove it with a vegetable peeler.

VANILLA

Do not substitute vanilla extract for vanilla beans in this combination. In addition to the complementary aromatic qualities it offers, vanilla's visual aspect—its little black specks—is important. They add a level of intrigue and surprise in a savory dish and reveal the use of high-quality ingredients. The rich flavor provided by the whole bean intensifies the preparations.

LAVENDER

Use this herb, whether fresh or dried, sparingly. Its flavor can be strong and sometimes bitter. When used with a light hand, it can be just the addition of minty floral flavor that makes a dish spectacular.

"SAVORY, FLORAL, AND FRAGRANT" RECIPE

Fragrant and Sweet Roasted Parsnips

These parsnips could accompany any fish or meat dish and may be served hot or at room temperature. For a more interesting presentation, cut the parsnips the day before, cover with water, and refrigerate overnight. They will curl up and look less stick-like. But don't worry, this is for aesthetics only—it won't affect the flavor.

2 cloves garlic, finely chopped

2 tablespoons (30 ml) olive oil

1 tablespoon (20 g) honey

½ **vanilla bean**, split lengthwise
   and seeds scraped (reserve pod
   for future use)

½ teaspoon salt

½ teaspoon fresh ground pepper

5 medium **parsnips**, peeled and
   cut into sticks

1 teaspoon (1 g) **dried lavender leaves**

Preheat oven to 400°F (200°C, or gas mark 6). In a large bowl combine garlic, olive oil, honey, vanilla seeds, salt, and pepper. Add parsnip sticks (or curls) and toss to coat. Place in a roasting pan and cook for 30 minutes or until tender and golden. Sprinkle with lavender and toss. Serve warm or at room temperature.

Yield: 4 servings

Available year-round, parsnips are best in the fall. Use this combination with other seasonal favorites such as apples, pears, and squash.

"RED-HOT INNOVATION"

CHILE PEPPER + COCONUT + TOMATO

This trio of flavors can be spicy, sweet, creamy, or all of the above. The ingredients' different forms, as well as different cooking techniques, affect the outcome of your dish. Do you want a spicy, lively legume side dish? Try the application recipe with lentils. Are you looking for light, seasonal salad? How about a roasted chile-coconut heirloom tomato salad? Coconut prawns with a chile-tomato dipping sauce offer a fun combination of the ingredients as well. Ethnic cuisine, such as Indian cooking, or those more focused on maintaining the purity of the ingredients, such as California cuisine, can guide these flavors.

CHILE PEPPER

Fresh chiles are most potent when eaten raw but soften considerably when roasted. Dried chiles are readily available and are rehydrated upon use. For flavorful seasoning, use powders. Use a mild pepper such as poblano for a mild heat, jalapeño for more heat, and habañero to blow the lid off of a dish. To add a bit of spice, mince a fresh chile pepper and cook it with other aromatic vegetables such as onion and garlic.

COCONUT

Fresh coconut is not as sweet as the processed variety. Coconut milk, sold in cans in most grocery stores, is a traditional base for Thai cooking. To increase the coconut impact in this combination, use coconut milk as a sauce component, as well as flaked coconut as a garnish.

TOMATO

Eat raw and slice tomatoes in salads. Sauté them for stir fries. Purée them into sauces for pasta, meat, and fish or oven-roast them to deepen their flavor. Sun-dry them to bring out their natural sugars. To use tomato as an accent flavor, use chopped sun-dried tomatoes with roasted peppers and shredded coconut for a flavorful tapenade for grilled meats and vegetables. Bump up the tomato by making a sauce with a chile accent.

"RED-HOT INNOVATION" RECIPE

Chile-Spiked Lentils

Serve this great vegetarian dish warm or at room temperature.

2 tablespoons (30 ml) vegetable oil (such as canola)

1 small onion, chopped

1 inch (2.5 cm) fresh ginger, peeled and minced

1 teaspoon (2 g) cumin seeds

2 **whole dried chiles** (such as chile de arbol)

1 **fresh hot green chile** (such as jalapeño), cut in thin slices crosswise

2 tablespoons (10 g) **unsweetened shredded coconut**

2 small **tomatoes**, chopped

1 cup (192 g) lentils, simmered in water until tender and drained

Juice of 1 lime

¼ cup (4 g) chopped fresh cilantro

Salt and pepper

Heat oil over medium heat in a large saucepan. Add onion and cook until just tender. Add ginger and cumin and stir, cooking until onions become translucent and cumin browns. Add dried chiles, cooking for 1 minute before adding fresh chiles. Cook for an additional 1 minute and stir in the coconut until it browns.

Add chopped tomatoes, lentils, and lime juice. Bring to a simmer and then remove from heat. Stir in cilantro and season with salt and pepper. Serve warm or at room temperature. You may refrigerate this dish for several days before serving.

Yield: 4 servings

Make this less spicy by reducing the amount of chiles used or omitting the fresh green chile pepper altogether.

"THE SALACIOUS PARTY"

PEPPERCORN + CORIANDER + CARDAMOM

This blending of spices produces an exotic, flavor-packed taste made more mysterious in a traditionally sweet pastry. The slap of heat produced by freshly ground pepper bears little resemblance to the lackluster preground version. The nutty, citrusy coriander and the highly aromatic, floral cardamom provide an intriguing blend.

In the application recipe, these spices are folded into sweet, buttery dough to produce a savory shortbread. Serve with frozen yogurt, cheeses, or fresh figs. Also, add these spices to sauces, soups, and seasonings for meat and vegetables.

PEPPERCORN

Black peppercorns, the peppercorns most readily available, are a strong flavor choice for this application. Other interesting options include pink peppercorns, which have a delicate, fragrant flavor and are aesthetically pleasing, or white peppercorns, the same plant as the black but fully ripened and with a more complex flavor. For maximum flavor, buy the peppercorns whole and grind them yourself.

CORIANDER

The coriander seed has a nutty flavor accentuated by roasting. Buying whole seeds, roasting them, and grinding them gets the maximum flavor possible. Ground coriander is acceptable; just be sure to use it soon after buying, as its flavor diminishes over time.

CARDAMOM

Sold as whole pod or ground, this intensely aromatic seasoning is common in Middle Eastern, Indian, and Scandinavian cuisines. Its multifaceted flavor gives it extensive uses in both savory and sweet applications. For this recipe, buy whole pods and toast and freshly grind them.

"THE SALACIOUS PARTY" RECIPE

Savory Shortbread

Normally, butter and sugar are shortbread's reliable headliners, but in this combination, the spice blend waits in the wings as the understudy. Be sure to use real butter; margarine or shortening just don't produce the same flavor.

1½ (2 g) teaspoons whole **cardamom seeds**

1 teaspoon (2 g) whole **peppercorns**

1 teaspoon (2 g) whole **coriander seeds**

1¼ cups (275 g) unsalted butter, softened, plus more for pan

½ cup (100 g) sugar

1 teaspoon (5 ml) vanilla extract

3 cups (375 g) all-purpose flour

¼ teaspoon salt

Shortbread freezes well, making it a nice do-ahead project for the holidays. Freeze it either as dough or after it has been cooked and cut.

Grease a 10-inch (25 cm) square baking pan and line with parchment paper. Preheat oven to 350°F (180°C, or gas mark 4).

In a small skillet, place spices over medium heat and toast until fragrant and slightly browned, 2 to 3 minutes. Cool and finely grind in a spice grinder or cleaned-out coffee-bean grinder.

Beat butter in a standing mixer with a paddle attachment for 1 minute. Add sugar and vanilla and beat until very light in color and fluffy, 10 minutes. Add flour, salt, and ground spices and beat on lowest speed until just combined. Use a rubber spatula to scrape bowl and paddle to incorporate.

Press the dough into the prepared pan, patting with floured hands to make the top even. Use a fork to pierce the surface at ½-inch (1¼ cm) intervals. Bake for 20 minutes until lightly browned and firm. Let cool in the pan for 20 minutes.

Remove and cut before it is completely cool. Use a ruler to make uniform 2-inch (5 cm) pieces or cut on the diagonal for diamond shapes. Serve immediately or keep tightly covered for several days.

Yield: About 30 pieces

"GILLIGAN'S FAVORITE CHOCOLATE"

CHOCOLATE + GINGER + NUT

Chocolate and nuts naturally go together, but giving them the flavor-packed thrust of ginger makes for a stimulating bite. Chocolate softens ginger's zing, and the nuts provide a pleasing crunch to balance the creaminess. The combination is balanced and interesting without being overwhelming—a lovely end to a meal.

In the application, chocolate flavor dominates. To turn it on its head, dip pieces of crystallized ginger in tempered chocolate, sprinkle with ground nuts, and allow to dry. Another possibility: Prepare a quick bread adding cocoa powder, ground ginger, and nuts to the mix.

CHOCOLATE

Chocolate ranges from unsweetened and semisweet to milk chocolate, with varying amounts of sugar versus cocoa creating each and affecting the flavor of the finished dish. To sweeten this combination and divert attention from the ginger, use milk chocolate. Otherwise, for the most intense flavor, stick with a high-quality bittersweet chocolate with a minimum of 50 percent cocoa mass.

GINGER

Its juice is potent, and the fresh root packs a flavor punch as well as texture. Cooking ginger mellows its impact and allows it to incorporate with other flavor profiles. Pickling ginger retains its crunchy texture, while candying or crystallizing it gives it a chewy, sweet taste.

NUT

Toasting and roasting nuts intensifies their flavors. In this combination, let the overall flavor profile affect your choice of nut—a bold taste can handle a stronger-flavored nut such as a walnut, while a milder dish tastes better with a mild macadamia or pecan.

"GILLIGAN'S FAVORITE CHOCOLATE" RECIPE

Triple-Threat Truffles

Truffles seem a fancy dessert for the home baker, but they are actually quite easy to make. This recipe simplifies them even further by rolling them in chopped macadamia nuts instead of dipping them in tempered chocolate. Rather than making brownies for your next potluck, make these instead—they will disappear!

2 inches (5 cm) **fresh ginger**

½ teaspoon fresh lemon juice

1 pound (455 g) **semisweet chocolate**, chopped

1¾ cups (410 ml) heavy whipping cream

6 tablespoons (50 g) finely chopped **crystallized ginger**, divided

1 cup (135 g) toasted **macadamia nuts**, finely chopped

Peel and grate fresh ginger. Squeeze to extract 2 tablespoons (30 ml) of juice. Mix with lemon juice and set aside.

Place chocolate in a medium-size bowl. In a saucepan over medium heat, bring cream to a simmer. Pour cream over chocolate and allow to sit for 1 minute. Whisk together until smooth. Add 2 tablespoons (17 g) crystallized ginger and ginger juice; stir until incorporated. Place in refrigerator and allow chocolate mixture to cool until firm.

While chocolate cools, combine remaining crystallized ginger and chopped macadamia nuts. Place in a shallow bowl.

When chocolate is firm, scoop 1-inch (2.5 cm) balls with a mini ice cream scoop and drop into ginger-nut mixture. Roll to coat all over and repeat. Store tightly covered in the refrigerator. Remove 1 hour before serving. Truffles are best eaten at room temperature.

Yield: 30 truffles

Truffles make a nice homemade gift, and a little effort in presentation yields a lot of "Wow." Place them in paper candy cups tucked into small boxes. Just be sure to tell the recipient to keep the gift refrigerated.

"SWEET AND NAUGHTY"

SODA POP + LIQUOR + CREAM

The fizzy, sugary soda pop has been a longtime favorite for a cocktail mixer. Adding an element of cream softens the sweetness and alcohol flavors and gives richness that can easily translate into desserts.

The application recipe prepares a fun whiskey-and-Coke ice cream. Another option: Use your favorite soft drink and liquor for a float—root beer with vanilla vodka and ice cream or cream soda with golden rum and coffee ice cream.

SODA POP

The application recipe uses cola syrup, a good way to use cola flavor in custards without the carbonation of soda pop. You also can make your own soda by adding seltzer water to it. Try root beer, cola, cream, ginger, or lemon-lime soda pop flavors. There are also gourmet sodas out there—the brand Dry Soda makes our favorites, which include flavors such as rhubarb, lavender, and kumquat.

LIQUOR

Look to your favorite mixed drink to guide you in choosing a liquor for this combination. Rum or bourbon go great with cola, Seagram's 7 is a classic with lemon-lime, and vodka is great with root beer or fruit-based sodas. Adding cassis (black currant–flavored liqueur) or Chambord (raspberry-flavored liqueur) results in an interesting take on a cherry cola.

CREAM

Use a heavy cream as a splash in a drink or let cream lead in the combination by making an ice cream, panna cotta, or sherbet. Don't be tempted to use the skim milk hanging around in your refrigerator—the dairy here needs fat to create richness for contrast with the other flavors.

"SWEET AND NAUGHTY" RECIPE

Jack & Coke Ice Cream

This fun ice cream makes a great float in Coke, of course!
The cola syrup is available at drugstores or online.

Special equipment required

6 egg yolks
²/₃ cup (160 ml) **cola syrup**, divided
2 cups (475 ml) **cream**
¼ cup (60 ml) **Jack Daniel's whiskey**

Fill a large bowl of water for an ice bath and set aside. In a medium-size bowl, whisk the egg yolks together with 1 tablespoon (15 ml) cola syrup until blended well.

Pour the cream and the remaining cola syrup into a large saucepan over medium heat. Bring to a simmer. Using a small ladle, add ¼ cup (60 ml) hot cream and syrup to the egg mixture and whisk together. Repeat with another ¼ cup (60 ml) of the hot cream and syrup mixture. In a steady stream, add the egg mixture back into the hot cream and syrup in the saucepan, whisking constantly. Continue to cook, using a rubber spatula occasionally to scrape the pan's edges and bottom.

When the mixture thickens to the consistency of a thick cream, it is done cooking. Pass it through a fine mesh strainer into a bowl to place on top of the ice bath. Allow to cool completely, stirring occasionally. Twist the ice cream according to manufacturer's instructions for your machine. Just before the ice cream finishes churning, add the whiskey. Store covered in the freezer for at least 2 hours prior to serving.

Yield: 3 cups (420 g)

For a variation, use rum instead of whiskey and add candied lime rind for a Cuba libre ice cream.

"A HOT POP OF FLAVOR"

CARAMEL + CORN + CHILE PEPPER

Caramel's sweet intensity matches wonderfully with corn's inherent sweetness. Bringing the spicy chile pepper into the group enlivens the mix. Vary the combination by changing how you use the caramel—have a lighter hand for savory dishes or a heavier hand for desserts. Vary the intensity of spice from the chiles by using raw for the most heat or sautéing or roasting to mellow.

The application recipe uses the popcorn kernels to create a dangerously addictive sweet-and-spicy caramel corn. Take the combination to the savory side with a batch of vegetarian chili and caramel butter topped cornbread or grill corn on the cob with minced roasted jalapeños and brush with caramel sauce.

CARAMEL

Though making caramel is not complicated (just boil sugar and water), it does require both a careful eye because it burns easily and a steady hand because it is molten hot and can burn you. For a butterscotch flavor variation, use brown sugar.

CORN

Take kernels straight off of the cob and serve them raw for a crunchy bite. Cooking methods (boiling, grilling, sautéing), both for the whole cob or for kernels only, soften the taste and texture. Colorado Olathe corn is a deliciously sweet variety worth seeking. The application recipe uses popcorn. While a range of gourmet, colored popcorn kernels exist, yellow generally pops the biggest and fluffiest.

CHILE PEPPER

To evenly distribute the heat of chile directly into a hot caramel, use dried chile flakes or ground chile pepper (such as chipotle). When grilling corn, adding canned chipotle chile in adobo makes it easy to both brush the sauce on the cobs and mince the peppers to add to a caramel sauce. They are spicy—start with a small amount and taste to adjust to your preference.

"A HOT POP OF FLAVOR" RECIPE

Knockout Caramel Corn

Pop the corn using whichever method you prefer. Be sure to keep it in an airtight container and store for up to a week (not that it will last that long).

2 tablespoons (30 ml) canola oil

½ cup (104 g) **popcorn kernels**

½ cup (56 g) cashews, toasted

2 cups (400 g) **sugar**

1½ tablespoons (27 g) salt

½ cup (120 ml) water

3 tablespoons (42 g) unsalted butter

1 teaspoon (4.6 g) baking soda

1 tablespoon (3.6 g) **dried
    chile flakes**

Line a sheet pan with parchment paper. In a large pot, heat the oil over high heat. Add the popcorn and cover with a lid. Cook until it begins to pop, shaking frequently. Continue cooking until the popping slows down to 3 seconds between pops. Transfer to a bowl, remove unpopped kernels, and toss popped kernels with cashews.

In a large, heavy saucepan, combine the sugar and salt. Pour water over the mixture, add the butter, and swirl the pan to ensure all the sugar is wet. Bring to a boil over high heat. Let sugar mixture boil until it starts to brown around the edges. Swirl the pan, but don't stir to distribute the heat, and continue to boil until sugar turns golden brown throughout, creating caramel.

Remove from heat and add the baking soda and chile flakes, being careful because the caramel will bubble up. Use a rubber spatula to combine the ingredients and add the popped corn and cashews to the pot. Working quickly, toss the popcorn with the caramel until well coated. Spread out on the prepared sheet pan. When cool, break into small pieces and serve.

Yield: 4 servings

Potential variations incorporate different types of nuts (pistachios, pecans, or pine nuts) or use a ground chile pepper (such as chipotle) as a substitute for the chile flakes.

CHAPTER 2

Unexpected Pleasures

Who doesn't want to experience pleasure? We hope for it, pursue it, and strive for it. We commit large amounts of our time, money, and effort to gaining it, if even for a fleeting moment. Yet there is nothing quite like being caught off guard by its pursuit. Opening your mouth to a burst of new flavor can have the same breathtaking effect as opening your front door to a huge rainbow.

Expectations can influence the perception of pleasure. One way to keep yourself more open to unexpected pleasures is to examine what you expect. Having negative or neutral expectations about something and being wonderfully surprised by it offers the best kind of treat.

What food hang-ups do you need to shed? Maybe when you were a kid, your older sister held you down and forced you to eat cocktail olives and you have never been able to go near them since. Or perhaps your own parents were not adventuresome with eating, and their conservative culinary habits seeped into yours. Why exactly don't you like radishes? Maybe you and your tastes have changed in the ten years since you last tried them, and it's time to taste them again.

In our restaurant, we have no written menus, so as we serve, we verbally tell our guests what they are eating. Occasionally, a guest will pull us aside and say something such as, "The fewer details you tell us, the better—my friends will enjoy it more if they don't know what they are eating." Then later, upon telling the unsuspecting diners that the dish they just gobbled up was frog legs, someone inevitably gasps, "Really? I had no idea!"

In this chapter, we pursue flavor combinations that may "sound weird" but make sense in an innovative new way. Weirdness, of course, is in the eye of the beholder, but these recipes generally fall into one of three categories: the ingredients themselves are not typically used together (chocolate with bacon, anyone?); the ingredients are under-appreciated ("I've never really liked anchovies, actually"); or the ingredients are prepared in a unconventional way (think grilled watermelon).

These recipes may force you to reflect on and change your expectations. You may have to convince yourself that trying goat's milk yogurt isn't really all that scary and that corn for dessert could be a good idea. Maybe you'll end up cleaning your plate, shaking your head, and happily saying, "Who knew?"

"DARK AND MYSTERIOUS"

CUMIN + CINNAMON + COCOA

Making your own spice blends is risky business. Tossing strong flavors together can create unappetizing tastes if you aren't careful. But this is a blend that works. The flavors play off of each other in a pleasing way. Toasting the spices deepens the profile, and grinding them together brings harmony. For the most punch, let cumin lead; use twice as much as cinnamon and cocoa for a dynamic taste. Be creative with this combination—try it as a rub for poultry or game meats, to season shrimp and seafood, or as an accent in demi-glace sauces.

CUMIN

Used to season both as whole seeds and ground, cumin often makes appearances in chili powder, curry powder, and Tex-Mex dishes. Do not use ground cumin for this recipe. Use whole cumin. To intensify the flavor, toast and grind the seeds shortly before use.

CINNAMON

This application calls for toasting a cinnamon stick and grinding it with the other spices to get the most flavor possible. Make cinnamon more prominent in this combination by going sweet, as in cinnamon whipped cream for a cup of Mexican hot chocolate with a sprinkle of ground cumin.

COCOA

In this application, cocoa powder adds a complex chocolate depth without the sweetness associated with chocolate. Make sure to use only unsweetened cocoa products and not hot chocolate mixes, which are sweetened and would not work with these flavors. Increase the amount of cocoa powder in a spice blend to create an interesting flavor addition for ice creams and cream sauces.

"DARK AND MYSTERIOUS" RECIPE

C-Spice Blend

This is quite the versatile spice blend to have on hand. Use it to season beef, pork, or poultry. For a refined presentation, sprinkle it around your serving plate for both visual and flavor interest.

2 pods cardamom

2 teaspoons (4 g) **cumin seeds** (not ground)

1 teaspoon (2 g) coriander seeds (not ground)

1 tablespoon (4 g) ground coffee

1 **stick cinnamon**

3 whole cloves

2 tablespoons (15 g) spicy chili powder (paprika or chipotle)

1 tablespoon (5 g) **unsweetened cocoa powder**

1 tablespoon (18 g) salt

2 tablespoons (30 g) packed brown sugar

Gently toast the cardamom, cumin, coriander, coffee, cinnamon, cloves, and chili powder in a dry skillet (without oil) over medium heat. Cook 4 to 5 minute or until spices become fragrant and begin to smoke.

Add cocoa powder, salt, and brown sugar and remove from heat. Allow to cool completely and then grind in a spice mill or clean coffee-bean grinder. Keep in a tightly covered container until ready to use.

Yield: ½ cup (50 g)

A spice mill or designated coffee-bean grinder is a good investment for flavorful cooking. Store-bought ground spices do not stay fresh indefinitely. Grinding your own spices in small batches ensures the maximum flavor.

"SWEET AND SALTY GONE SEDUCTIVE"

FIG + APPLE + ANCHOVY

In this unusual combination, sweet, crispy, and salty combine to produce pleasing results. The soft, jammy fresh fig is a seasonal fruit partner with the crunchy apple. Both play host to the salty intensity of anchovy and combine in a savory, complex treat. For versatility, use the fig and apple raw or cooked.

This application recipe prepares a crostini with a fig-anchovy spread and sliced apple. But try this combination in a tomato-anchovy sauce for swordfish topped with apple-fig chutney. Other flavors to incorporate into this combination include capers, olives, and cheeses such as Parmesan.

FIG

Figs enjoy a short season from late summer throughout the fall months. They are also dried and are effectively rehydrated for use in sauces and vinaigrettes or reduced to a syrup glaze. For fresh varieties and flavor versatility, look for the dark purple Black Mission or the light green Kadota. Simmer dried figs in water or wine to soften before using. Fresh varieties work best for this combination.

APPLE

Use raw apples to get the greatest texture and crisp flavor. For raw applications in this combination, seek out Pink Lady or Gala apples. Cooking softens the intensity of the apple's flavor and texture. Tart Granny Smith apples balance the sugars that increase as the apples cook. Use apple juice to reconstitute dried figs, if using.

ANCHOVY

The most readily available anchovies, which come packed in olive oil, can work nicely for this combination, especially when puréeing, as in the application recipe. To use anchovy more prominently, look for the delightfully mild Italian white variety or use the whole fresh fish, great with something like apple-fig chutney.

"SWEET AND SALTY GONE SEDUCTIVE" RECIPE

Autumn Anchovy Crostini

These crostini make a great appetizer to pass at a cocktail party. Or slice your baguette on the bias for a large piece and serve with greens for a salad.

1 baguette, sliced thinly into rounds
Olive oil, for drizzling
Salt and pepper
6 **fresh figs** (preferably Black Mission), sliced in half
1 clove garlic, peeled
3 **anchovies** packed in oil (preferably Italian white)
1 tablespoon (9 g) small capers
1 teaspoon (4 g) Dijon mustard
1 **apple** (such as Pink Lady), quartered, cored, and cut into thin slices
Shavings of Parmesan cheese (preferably Parmigiano-Reggiano)

Preheat oven to 350°F (180°C, or gas mark 4).

Drizzle slices of bread with olive oil and sprinkle with salt and pepper. Bake until lightly browned and crisp.

In a food processor, pulse figs, garlic, anchovies, capers, and mustard until just blended. Season with pepper.

Spread fig purée on crostini, top with 2 or 3 apple slices and shavings of cheese, and serve immediately.

Yield: 12 small or 4 large crostini

Seasonally, this is a fall dish, when figs and apples are both at their peak. You may use dried figs for this if you don't have access to fresh; just be sure to cook in some liquid (such as apple juice) to rehydrate before puréeing.

"BOLD AND BOISTEROUS"

TOMATO + BROWN SUGAR + COFFEE

These three ingredients rarely end up on the same plate. Yet when combined, they result in a bold taste sensation. The tomato brings in the deep sweetness of the brown sugar, which interacts with the tomato's own natural sweetness. Coffee is the surprise, somehow deepening this sweetened tomato combination.

The application recipe produces a tangy barbecue sauce by cooking the combination together with spices. For a different twist, try coffee-braised short ribs with tomato brown sugar glaze. Serve this combination with beef or pork ribs, chicken, or pork.

TOMATO

Cooking applications such as roasting intensify a tomato's flavor, while drying brings out its natural sugars. Cooking the tomato with the other ingredients helps to incorporate its flavors and smooth it out. Because you will likely cook the tomato in this application, rely on high-quality canned products for consistent flavor, especially during off-season times of year.

BROWN SUGAR

Brown sugar provides a richness and depth of flavor in baked goods, warm dessert sauces, and sweetened marinades and vinaigrettes that white sugar simply can't match. If you need a quick substitute for brown sugar, combine 1 tablespoon (20 g) of molasses to ½ cup (100 g) of white sugar. Use the mixture in equal proportion to the required amount of brown sugar.

COFFEE

How coffee beans are roasted affects a coffee's flavor—from light and slightly bitter to dark and smooth. When using in cooking, brew a strong batch—even stronger than you may like to drink. To appreciate it among these other flavors, the coffee needs its own boldness.

"BOLD AND BOISTEROUS" RECIPE

Barbecue Sauce for Any Occasion

*Think beyond the grill for this barbecue sauce—as a sauce for pizza,
a substitute for ketchup, a spread for meat loaf, or to cook with tofu.*

2 tablespoons (30 ml) olive oil

1 medium onion, finely chopped

8 cloves garlic, finely chopped

1 small poblano pepper, seeded and finely minced

¾ cup (170 g) packed **brown sugar**

1 tablespoon (20 g) light molasses

3 tablespoons (3 g) chopped fresh cilantro

1 teaspoon (2.5 g) ground cumin

1 can (28 ounces, or 820 g) **crushed tomatoes in purée**

1 cup (235 ml) chicken broth

1 cup (235 ml) strong **coffee**

Salt and pepper

Heat oil in a large saucepan and cook onion, garlic, and poblano pepper until tender. Add brown sugar, molasses, cilantro, and cumin. Stir together and cook until brown sugar dissolves. Add tomatoes, broth, and coffee and stir until combined. Let simmer until thickened and reduced in quantity to 1 quart (1 kg). Adjust seasoning with salt and pepper. Let cool, cover, and keep refrigerated for up to 1 week.

Yield: 1 quart (1 kg)

Make a large batch of this unique sauce and put into canning jars. This makes a great host gift for your next party and is more personal than a bottle of wine.

"PUCKER UP FOR A KISS"

WATERMELON + CHEESE + VINEGAR

Watermelon's cool, crisp flavor and light sweetness are a surprising contrast to creamy cheese. The acidic vinegar accentuates the flavors, drawing out a dynamic combination.

Our application recipe uses a rare cooking technique for the watermelon—grilling it to slightly caramelize the flesh. It is then topped with feta cheese and balsamic vinegar. Or for a more straightforward recipe, toss the components together for salad topping, varying the cheese and vinegar selections to create different tastes.

WATERMELON

Search out the yellow and orange varieties and use them with the traditional red for a colorful presentation. Watermelon is usually eaten raw or puréed into drinks, not frequently cooked, but the following recipe calls for grilling thick slices. Save what you normally toss into the compost heap and pickle the rind. Or scrape your cutting board full of watermelon juice—what's left after slicing—into a bowl with a light vinegar and oil to dress a salad.

CHEESE

Soft, mild cheeses such as goat cheese, feta, and ricotta taste great with watermelon because their creaminess and softness contrast nicely with the water-drenched melon. To create a more flavorful cheese presence in the combination, try a creamy blue cheese. Avoid hard cheeses such as Parmesan that don't provide the pleasing texture contrast.

VINEGAR

For a light vinegar impact, choose a Champagne or white wine vinegar to use as a dressing. To make a stronger effect, use balsamic vinegar, particularly an aged one (which can be expensive but is worth it because you use it sparingly). Balsamic vinegar's sweet, mellow, full flavor can transform a dish like this one.

"PUCKER UP FOR A KISS" RECIPE

Succulent Grilled Watermelon

Serve these grilled watermelon slices over a bed of greens for a salad, adding grilled chicken or shrimp to make it a meal.

1 small **seedless watermelon**, cut into
   1-inch (2.5 cm) slices, rind removed
1 tablespoon (15 ml) olive oil
1 tablespoon (20 g) honey
Salt
1 cup (150 g) crumbled **feta cheese**
2 tablespoons (30 ml) **aged balsamic
   vinegar**

Heat grill to high. Place watermelon slices on paper towels to drain excess liquid.

In a small bowl, combine oil and honey. Brush the mixture on both sides of the watermelon, sprinkle with salt, and place on the grill. Cook until lightly browned, 1 to 2 minutes on each side.

Remove from heat and sprinkle with crumbled feta. Drizzle vinegar over the top and serve immediately.

Yield: 4 servings

This is a summer dish to be enjoyed at the height of melon season.

"INTERPLAY OF INTENSITY"

COCOA + RED WINE + MINT

This is a richly complex combination, with the deep intensity of cocoa powder, the bold burst of red wine, and the bright lift of fresh mint. Cocoa and mint often get paired (think minty hot chocolate), as do cocoa and red wine for baking applications (think cocoa merlot cake), but it's rare to find them all grouped together.

In the application recipe, they become a braising liquid for duck (also great for chicken or other poultry), with slow, long cooking that softens and incorporates the flavors. To preserve the mint's zing, add it at the very end. These bold flavors work well with rich meats such as duck, lamb, and squab.

COCOA

Not to be confused with the chocolate normally found in finished candy bars, cocoa is a product of the processed leftover solids from making chocolate. It is an unsweetened powder used in baking, desserts, and beverages. In this application, it adds a complex chocolate depth minus the sweetness typically associated with chocolate. Use cocoa as an underlying accent—as in this combination—or more pronounced, as in desserts.

RED WINE

Don't substitute white wine for this combination, as it will not provide the flavor profile suitable with the cocoa. Go for a full-bodied Cabernet Sauvignon or Zinfandel (not pink!) for maximum impact. In sweeter applications, consider using a port wine for its increased sweetness and flavor concentration.

MINT

Mint provides an element of surprise to this dish, but use it with a delicate hand. Consider it the accent flavor, not a dominant taste, and incorporate accordingly—a sprinkle of chopped mint over a finished dish or combined with other herbs to give a bright note to an otherwise luxuriously rich flavor combination.

Braised Duck with a Lift

Braised dishes are great for entertaining because most of the work takes place at the front end. Serve this with risotto or mashed potatoes for a satisfying meal.

4 duck legs and thighs (legs and thighs
   typically sold together as 1 piece)

Salt and pepper

4 sprigs fresh thyme

1 medium yellow onion, chopped

1 large carrot, peeled and chopped

4 cloves garlic, peeled and chopped

1 inch (2.5 cm) fresh ginger, peeled
   and chopped

1 tablespoon (8 g) all-purpose flour

2 tablespoons (10 g) **unsweetened
   cocoa powder**

2 tablespoons (12 g) chopped **fresh
   mint**, divided

1 cup (235 ml) **dry red wine**, such as
   Cabernet or Syrah

2 tablespoons (30 ml) Cointreau liqueur

Zest of 1 orange

½ orange, cut into 6 slices

2 cups (475 ml) water

Juice from ½ lemon

Preheat oven to 275°F (140°C, or gas mark 1). Season duck with salt and pepper. In a large skillet, sear the duck on both sides over medium-high heat. Place duck in a large oven-safe cooking pan and arrange thyme sprigs on top.

In the same skillet used for searing, add the onion, carrot, garlic, and ginger. Cook until softened, 2 to 3 minutes. Add the flour and cook for an additional 2 to 3 minutes.

Add the cooked vegetable mixture to the cooking pan, along with the cocoa powder, 1 tablespoon (6 g) mint, wine, Cointreau, zest, orange pieces, and water. Cover with lid or foil and bake for 1 to 1½ hours until the duck is very tender.

Remove the duck from the pan to a serving dish. Transfer the liquid and vegetables to a saucepan, bring to a simmer, and reduce slightly to a sauce consistency. Stir in remaining 1 tablespoon (6 g) mint and lemon juice and adjust seasoning with salt and pepper. Serve duck with sauce in shallow bowls.

Yield: 4 servings

"ASIAN FLIRTATION"

CHESTNUT + MISO + ORANGE

Sweet, delicate chestnuts make for an unusually tasty blend with the Asian-influenced combination of miso and orange. The earthy soy flavor of miso plays off of the sweet citrus, enhanced by the soft nuttiness of the chestnuts.

The application recipe suggests puréeing the chestnuts and serving them with poached fish in an orange-miso glaze. Other applications include roasted chestnuts with an orange-miso vinaigrette or as additions to rice or sweet potatoes. These flavors are well suited to seafood, winter squash, and poultry.

CHESTNUT

Roasting gives chestnuts a texture similar to baked potatoes, with a sweet, soft, delicate flavor. Boiling, steaming, grilling, and frying also work. Chestnuts taste delicious in savory applications such as stuffing or side dishes, in desserts soufflés, or in the candied form. Canned chestnuts are fine for purées, but rely on fresh for other applications. Fresh chestnuts are in season in the fall and winter months.

MISO

Miso, available in Asian or health food stores as a paste sold in tubs, jars, or vacuum-sealed packages, has an earthy, mild soy flavor. If you have a choice, go for an unpasteurized miso for its improved taste and health benefits.

ORANGE

Oranges reach their flavor peak during the winter months, which overlaps with the season for fresh chestnuts. For minimal flavor accent, use juice only. Add more intensity by bringing in zest and segments or reduce orange juice in a small sauté pan over heat to make a more concentrated flavor. For a tart flavor and dramatic color, seek out blood oranges when they are in season. To use citrus segments in your cooking and take advantage of their juicy fruit only, "supreme" them. To do this, cut off both ends and then the side peel of the citrus, removing all visible white pith with your knife. Then cut closely to the side of each segment, leaving pieces with no peel attached.

"ASIAN FLIRTATION" RECIPE

Sea Bass with Chestnut Purée and Miso-Orange Essence

The miso-orange sauce works as both a marinade for the fish and as a sauce for the finished dish. Prepare the components in advance, giving the miso-orange a chance to infuse the fish. Throw in a preheated oven, warm up the chestnut purée, and dinner will be ready in less than fifteen minutes.

1 tablespoon (15 ml) vegetable oil

2 cloves garlic, minced

2 inches (5 cm) fresh ginger, peeled and minced

2 cups (475 ml) **orange juice** (preferably fresh squeezed)

2 tablespoons (32 g) **miso paste**

1 tablespoon (15 g) packed brown sugar

1 tablespoon (15 ml) soy sauce

1 tablespoon (15 ml) rice vinegar

1 tablespoon (15 ml) mirin (cooking sake)

Salt and pepper

4 pieces sea bass, 6 to 8 ounces (170 to 225 g) each

2 cups (400 g) **unsweetened chestnut purée** (canned)

¼ cup (55 g) unsalted butter

½ cup (120 ml) carrot juice

⅛ teaspoon sesame oil

2 **oranges**, supremed (page 121)

Salmon also works nicely with these flavors, and using blood oranges as the segments in the sauce makes for a nice presentation.

In a skillet over medium heat, heat oil and sauté garlic and ginger until softened. Add orange juice, increase heat, and bring to a simmer. Allow it to reduce by one-quarter of its volume.

Lower heat and add miso, brown sugar, soy sauce, vinegar, and mirin, whisking to combine. Cook for 5 minutes, season with salt and pepper, and remove from heat. Allow to cool before using as a marinade.

Place fish in a shallow pan and use ¼ cup (60 ml) of the miso-orange sauce to pour over all sides of the fish. Cover and refrigerate for 4 hours. Set aside remaining sauce.

Meanwhile, in a saucepan, stir the chestnut purée with the butter, carrot juice, and sesame oil. Whisk until the butter melts. If needed, add small amounts of water until the mixture reaches the consistency of whipped mashed potatoes. Season with salt and pepper and set aside.

Preheat oven to 350°F (180°C, or gas mark 4). Place fish on parchment paper and cook for 12 minutes (for fish 1- to 1½-inches [2.5 to 4 cm] thick). Do not overcook. Fish is done when it is firm yet still barely translucent on the inside.

While fish is in the oven, rewarm remaining miso-orange sauce on the stovetop, turn off heat, and add orange segments. Warm the chestnut purée in the microwave or on the stovetop. Serve fish on warm chestnut purée topped with miso-orange sauce and segments.

Yield: 4 servings

"OPPOSITES ATTRACT"

MUSHROOM + ROSE + LAVENDER

Here is a combination of contrasts—the earthy mushroom with the floral rose and lavender. It is a compelling interplay of flavors delicious to eat. The success of this combination depends on moderating the use of the rose and lavender. Let the mushroom flavor lead and add the rose and lavender with a light hand.

The application uses a grilled portobello as its focus, with a rose-lavender beurre blanc. For another variation, incorporate a mild fish, which would accommodate the divergent flavors nicely.

MUSHROOM

The meaty portobello is often a vegetarian alternative, regarded for its earthy flavor. And because of its large size, it is also one of the best mushrooms to grill. Its high water content helps the mushroom grill well, keeping it tender and intensifying its flavor. For a simple grill, brush with olive oil, salt, and pepper.

ROSE

The flavor of rose comes to food primarily through rose water, a clear liquid distilled from rose petals. Highly aromatic, it is typically paired with desserts. When cooking with rose water, start off with a small amount and taste, as the amount of concentration can vary by brand. Look for rose water in Middle Eastern markets.

LAVENDER

Cook with fresh or dried lavender. If using dried, be sure to buy a food-grade product and not one intended for potpourri, which could have additional oils not fit for consumption. Lavender's strong flavor goes nicely with other strong flavors (such as lamb) or to accent other sweet flavors (such as shortbread). The essential oil is quite potent, too strong for many cooking applications, but if you use it, be sure to count out the drops as specified in the recipe or risk having your dish taste like potpourri.

"OPPOSITES ATTRACT" RECIPE

Grilled Portobellos with Floral Beurre Blanc

Adding rose water and lavender to a traditional beurre blanc makes this dish memorable. Serve it over grilled portobello mushrooms and add rice for a meatless meal.

Beurre Blanc

⅓ cup (80 ml) dry white wine (such as
    Sauvignon Blanc)

1 shallot, thinly sliced

1 teaspoon (1 g) **dried lavender buds**

½ teaspoon whole peppercorns

1 cup (225 g) cold unsalted butter,
    cut into small cubes

½ cup (120 ml) **rose water**

Mushrooms

⅓ cup (80 ml) olive oil

1 tablespoon (15 ml) balsamic vinegar

1 tablespoon (15 g) packed brown sugar

1 clove garlic, minced

Salt and pepper

4 large **Portobello mushrooms**,
    stems trimmed

A separated beurre blanc
is called "broken," or no
longer emulsified. To bring
it back together, drop in
drips of cold water while
whisking it off the heat. Use
it immediately once brought
back together.

To prepare beurre blanc: In a skillet, cook the wine, shallot, lavender, and peppercorns until reduced to 2 tablespoons (30 ml) of liquid. Remove from heat and add ⅓ cup (75 g) of the cold butter, whisking to incorporate. Return to low heat and continue to whisk until the butter mostly melts. Continue to add the butter, a few cubes at a time, until incorporated. Repeat until all the butter has been added.

Strain through a fine mesh strainer and return to a small saucepan. Off heat, add the rose water 2 to 3 tablespoons (30 to 45 ml) at a time, tasting after each addition. Look for a clear but not overwhelming rose flavor. Fill a bowl with warm water and place saucepan in it to keep the sauce warm while grilling mushrooms.

To prepare mushrooms: Heat a charcoal or gas grill. Mix oil, vinegar, brown sugar, garlic, salt, and pepper to taste in a small bowl. Brush liberally on both sides of the mushrooms and grill on one side for 5 minutes. Turn, brush with more oil, and grill for an additional 5 minutes or until tender and browned all over.

Top grilled mushrooms with beurre blanc and serve immediately.

Yield: 4 servings

"TWISTED SUNSHINE"

CORN + CHILE PEPPER + ROSEMARY

There's no summer flavor like the down-home taste of fresh corn—unless, of course, you combine that flavor with chiles and rosemary. This combination has all of the elements for balanced, interesting flavor: the sweet corn, the spicy chile, and the herbal rosemary. Leave both the corn and chile peppers raw or cook them for endless variations—grilled corn on the cob with jalapeño-rosemary butter, sautéed chunky salsa, creamy puréed soup, or savory cornbread. These flavors are also great for polenta, chicken, and salmon. The unexpected element is the application recipe, which creates an unusual and delicious sorbet that you may just have to try to believe.

CORN

Take kernels off of the cob and serve them raw for a crunchy bite. Corn also purées nicely for soups or for the following sorbet recipe. Corn is inherently sweet. Why not try corn crêpes with a chile-rosemary cream? Save the cobs to make a corn stock or run the kernels through a vegetable juicer for a concentrated corn juice to use for vinaigrettes and sauces.

CHILE PEPPER

Fresh chile peppers are most potent when eaten raw but soften considerably when roasted. Dried chiles are readily available and rehydrated upon use. For flavorful seasoning, use powders. Any form of these chiles can work in this combination, as well as raw in a salsa, sautéed in a soup, or to accent cornbread.

ROSEMARY

The fresh evergreen-like needles impart a woody taste to a wide variety of foods, from fruits and vegetables to meats and desserts. Dried rosemary contains much less flavor than fresh, and because you can find fresh rosemary year-round, there is no excuse to use the dried version!

"TWISTED SUNSHINE" RECIPE

Stone Cold Corn Sorbet

This savory flavor-packed sorbet is especially nice with another summer treat—tomatoes.Serve it as a side to tomato soup or treat it as dessert with cornmeal shortbread or caramel corn.

Special equipment required

5 tablespoons (70 g) unsalted butter,
    divided

1 sprig **rosemary**

4 ears **sweet corn**, kernels removed
    (reserve cobs for stock)

1 **red jalapeño pepper**, sliced, seeds
    removed, and chopped

½ large sweet yellow onion, diced

4 cloves garlic, chopped

Juice of 1 lime

¼ cup (60 ml) simple syrup (page 183)

1 cup (235 ml) corn or vegetable broth

2 tablespoons (30 ml) extra-virgin
    olive oil

1 tablespoon (15 ml) dry white wine
    (such as Chardonnay)

1 tablespoon (15 ml) Champagne vinegar

Salt and pepper

In a large skillet over medium-high heat, melt 1 tablespoon (14 g) butter and add rosemary, corn kernels, jalapeño, onion, and garlic. Cook until soft, 5 minutes. Let cool slightly, remove rosemary, and discard it. Transfer to a blender.

Add lime juice, simple syrup, broth, the remaining ¼ cup (55 g) butter, olive oil, wine, and vinegar. Blend until very smooth. Adjust seasoning with salt and pepper and allow to cool or use an ice bath to cool quickly. Add to an ice cream machine and process according to manufacturer's instructions. Keep covered in the freezer until use.

Yield: 2 cups (300 g)

When serving sorbet or ice cream for entertaining, save yourself a step by prescooping. Place a parchment-covered tray in the freezer to chill and then scoop directly out of the ice cream machine, if the ice cream is hard enough. If you must pack sorbet into a container and freeze it before it's hard enough to scoop, scoop before people arrive and keep balls in the freezer. Then you can spend more time with your guests and less time wrestling with the sorbet.

"A TART ALTERNATIVE"

GOAT'S MILK + ROSEMARY + RHUBARB

This combination features some wonderfully complex flavors, creating an exotic but delicious taste. Goat's milk offers a tangy alternative to cow's milk, providing the opportunity to infuse the rich milk with the herbal rosemary. The addition of rhubarb into the combination wakes up the palate, with the vegetable's tartness and bright flavor.

The application recipe prepares a goat's milk–rosemary gelato with a rhubarb compote. Another alternative: Rhubarb cobbler with goat's milk–rosemary crème anglaise.

GOAT'S MILK

More complex in flavor than cow's milk, goat's milk is gaining popularity in the United States for those with allergies to cow's milk and for those who choose to avoid cattle byproducts (goats both eat less and occupy less grazing space than cows). Try it for its unique flavor—slightly sweet, with a sometimes-salty undertone. It's different enough to notice but familiar enough to enjoy. Also try goat yogurt and goat butter, available at health food stores along with goat's milk.

ROSEMARY

Rely on fresh rosemary (rather than dried) for a woody, pine flavor. Use sprigs to infuse the milk component by cooking together and removing the rosemary before serving. To increase rosemary's impact, leave the fresh chopped needles with the rhubarb in the finished dish.

RHUBARB

Rhubarb releases much of its moisture during cooking, which turns into a syrup when paired with sugar. It will break down into a pulp when cooked long enough, so if you want it to retain its shape, keep a close eye on it.

"A TART ALTERNATIVE" RECIPE

Luscious Gelato with Rhubarb Compote

Using goat's milk as the recipe calls for results in a deliciously different dairy experience. You can substitute cow's milk, but you'll lose the distinct flavor that goat's milk provides.

Special equipment required

Ice Cream

2 cups (475 ml) **goat's milk**

1 cup (235 ml) heavy cream

1 vanilla bean, split lengthwise
    and seeds scraped

½ cup (170 g) honey

2 sprigs **fresh rosemary**

4 egg yolks

Compote

2 cups (245 g) **diced rhubarb**

1½ cups (300 g) sugar

Juice and zest from 1 lemon

This is a great dessert to serve at a dinner party—it has impressive, unique flavors but is something you can prepare completely in advance (even to the lengths of prescooping your ice cream on a sheet pan in the freezer). Serve with a glass of sparkling Muscato or Champagne.

To prepare ice cream: In a large saucepan, combine goat's milk, cream, vanilla seeds and pod, honey, and rosemary. Bring to a simmer, stirring to dissolve the honey. Remove from heat, cover with plastic wrap, and let the flavors infuse for 20 minutes.

In a small bowl, whisk the egg yolks until thick and creamy. Bring the milk mixture back to a simmer; remove vanilla pod and rosemary, extracting as much liquid from them as possible. Add ¼ cup (60 ml) of the hot liquid to the eggs, whisking to combine. Repeat and then slowly add the egg mixture back into the hot milk in a steady stream, whisking constantly. Cook over low heat, alternating between using a rubber spatula and a whisk, until the gelato base thickens.

Pass through a fine mesh strainer and cool thoroughly over an ice bath or in the refrigerator, stirring frequently. Add to ice cream machine and process according to manufacturer's instructions. Use immediately or pack into a covered container and freeze for up to 1 week (gelato will be harder this way).

To prepare compote: Combine all ingredients in a saucepan and let it sit for 15 minutes. Bring to a simmer over medium heat and allow to cook, stirring often, until it thickens and the rhubarb breaks down.

Remove from heat and skim off any foam from the top. Cool, cover, and refrigerate for up to 1 week. Serve rhubarb compote with gelato.

Yield: 3½ cups (490 g)

"FOR GROWN-UPS ONLY"
CHOCOLATE + BACON + BROWN SUGAR

What happens when you take the two ingredients with the potential to make anything taste better (chocolate and bacon, of course) and pair them together? An oddly delicious combination of sweetness and smoke explodes in your mouth. Bring in rich brown sugar to both sweeten and deepen, and you have the ultimate in salty and sweet. This combination can translate into baking applications such as cookies and cupcakes or straightforward ideas such as chocolate-dipped candied bacon. Remember: It's always better with bacon!

CHOCOLATE
Vary the overall effect of this combination by changing what kind of chocolate you use. Milk and white chocolate are both quite sweet and highlight the brown sugar flavoring of the bacon. Semisweet or bittersweet chocolate allow the bacon flavor to shine through brightly. Use high-quality unsweetened cocoa powder for baking applications.

BACON
To make the most of the bacon in this combination, use the thickest-cut pieces you can find. Even if you end up chopping it up, you want substantial meaty morsels, not bacon dust. For an additional savory note, use peppered bacon; maple bacon will further develop the sweeter elements.

BROWN SUGAR
The added intensity of the molasses element in brown sugar makes a difference in the smoky bacon's flavor, so don't substitute white sugar in this recipe. If you need a quick substitute for brown sugar, combine 1 tablespoon (20 g) of molasses to ½ cup (100 g) of white sugar. Use the mixture in equal proportion to the required amount of brown sugar.

"FOR GROWN-UPS ONLY" RECIPE

Sweet Bacon Bark

This has nothing to do with sound your dog makes (although with the bacon in this dish, you may want to keep it out of your hound's reach). Chocolate barks are a great, simple way to make handmade chocolates that don't require special molds.

½ cup (115 g) packed **brown sugar**

7 strips **bacon**

Butter or spray oil, for greasing sheet pan

12 ounces (335 g) **semisweet chocolate**, chopped

Preheat oven to 350°F (180°C, or gas mark 4). Sprinkle brown sugar over strips of bacon and cook for 10 minutes. Remove and turn over pieces; cook for an additional 15 minutes or until crispy and dark golden. Let cool and then chop.

Butter a sheet pan or spray it with oil and line with parchment paper. Melt chocolate in the microwave, stirring at 30-second intervals, or melt over a double boiler over low heat.

Spread melted chocolate over prepared sheet pan, using an offset spatula to spread evenly to a ¼-inch (⅔ cm) thickness. Sprinkle bacon all over and allow to set at room temperature (or refrigerate, if pressed for time). Once hardened, remove from parchment and break into irregular pieces. Store covered in a cool place (preferably not the refrigerator) for up to 1 week.

Yield: 1 sheet tray (with varying-size pieces based on how the bark breaks)

You can put many different toppings on bark. Dried fruit, nuts, and cocoa nibs all work nicely.

"SWEET HEAT"

CHOCOLATE + VANILLA + CHILE PEPPER

It may seem counterintuitive to pair chocolate with vanilla, but that's only because of the misconception that vanilla equals no flavor or that it is the opposite of chocolate. Vanilla actually has rich aromatic qualities and a forward flavor used to provide depth to many dishes, both savory and sweet.

Here, chocolate is made more complex with the addition of vanilla, and then it's kicked in the pants by adding the spice of chiles. It is a surprising boost and makes you come back from each bite wanting more. Increase the chile element to make a unique mole sauce or cocoa salsa.

CHOCOLATE

Chocolate ranges from unsweetened and semisweet to milk chocolate, with varying amounts of sugar versus cocoa creating each and affecting the flavor of the finished dish. Use any variety in this combination; let the flavor of the finished dish guide you. (Here's a hint: milk for more sweetness, bittersweet for more chocolate intensity). Cocoa nibs, while not technically chocolate, also are made from pieces of the cocoa bean and impart a subtle nutty, chocolate flavor—a delicious possibility for this combination.

VANILLA

Do not substitute vanilla extract for vanilla beans in this application. Fresh vanilla beans produce the strongest vanilla flavor. Although extracts can be useful in baking applications, invest in whole beans to impart the most flavor possible.

CHILE PEPPER

Fresh chiles are most potent when eaten raw but soften considerably when roasted. Dried chiles are readily available and are rehydrated upon use. For flavorful seasoning, use powders. In this combination, if you choose to use a fresh chile pepper, try a poblano or chipotle for depth of flavor, as well as for heat.

"SWEET HEAT" RECIPE

Beggar's Purses with a Punch

These filled, fried "purses" are spiked with roasted chipotle, which results in a hot, spicy, chocolaty dessert in a beautiful presentation.

1 small **dried chipotle pepper**

½ cup (120 ml) espresso
(or strong coffee)

6 ounces (168 g) **bittersweet chocolate**, chopped

½ cup (112 g) unsalted butter, cut into 8 pieces

3 eggs

⅓ cup (67 g) sugar

2 **fresh vanilla beans**

8 large egg roll wraps (Melissa's brand works well)

Canola oil, for frying

When preparing these purses for the first time, give yourself a little extra product with which to work—you may want to purchase two packages of egg roll wraps (they freeze well if you don't use them).

Preheat oven to 300°F (150°C, or gas mark 2). Reconstitute pepper by placing it in a small, shallow pan and covering it with the espresso. Place in the oven and cook until the pepper softens and the liquid evaporates. Finely mince enough pepper to equal 2 tablespoons (18 g).

Melt chocolate and butter together over double boiler. Whisk together eggs and sugar in a medium-size bowl until the sugar dissolves. Scrape out the seeds from the vanilla beans and whisk into the eggs and sugar. Carefully cut scraped-out vanilla bean into 8 long pieces to be used to tie the purses. Add melted chocolate mixture to the eggs in a steady stream until fully incorporated. Fold in the minced chipotle, using a rubber spatula to evenly distribute. Wrap bowl with plastic wrap and refrigerate for 2 hours or until firm enough to scoop.

Place the egg roll wraps on a cutting board. Cut them in half lengthwise to make 2 rectangles. Cut one set of rectangles in half again to make 2 piles of squares. Cut the other set of rectangles into thirds, lengthwise, giving you 3 piles of long strips. Cover all dough with a damp towel when you aren't working with it so it doesn't dry out.

To make the first purse, lay out two strips in an X formation on your cutting board. Place a square sheet on the center of the X. Scoop out 2 spoonfuls of the chocolate mixture into the center of the square. Fold the corners of the square together and gather up the strips into the center. Tie with a strip of vanilla bean. Repeat with remaining purses.

Heat canola oil in a deep, heavy-bottomed saucepan to 380°F (193°C, or gas mark 5). Deep fry purses in oil until golden brown. Purses may be fried several hours in advance and reheated.

Yield: 8 purses

CHAPTER 3

Complex Creations

If the idea of complexity conjures images of writing a computer program or bringing peace to the Middle East, you can relax. In fact, relaxing with a cooking project is part of the idea here—don't expect to complete these recipes within thirty minutes, including commercials. Take off your watch, turn on some music, and enjoy the process.

We all know that greatness takes time—a seed transforming into a sprout then into a fruit-bearing plant does not happen overnight; the ripening of a single piece of fruit even requires patience—so why would we expect that we could prepare most foods in between checking emails? This chapter invites you tackle a more lengthy preparation and relax with it. You are reading this book because you have an interest in creating flavorful food, so act on that interest and don't let the complexity of these recipes daunt you.

The common thread uniting the recipes is something we all wish we had more of: time. In some, the few steps are rather straightforward. You may take extra care for an extra flourish—such as preparing a foam for figs in Roasted Figs with Crispy Bacon and Goat Cheese Foam (page 83) or an herbed oil for a drizzle on Sweet Parsnip Soup with Thai Basil Oil (page 87).

For others, the processes themselves take the time. Take Duck Prosciutto Salad (page 95), for example. After a simple technique of seasoning, curing, and wrapping the duck breast in cheesecloth, enter in the common ingredient of time—two weeks in this case—before you have prosciutto. Luckily, many of these time-consuming recipes are actually rather hands-off; much of the time, you don't have to be in the kitchen.

Sticking with some of these more time-consuming recipes will offer rich rewards. Going to the trouble of blanching lobsters, removing the meat, and preparing stock out of the shells results in rich bisque that you just can't replicate with a prepared product.

You can always let other people cook for you—at the drive-through, at a gourmet restaurant, or by going home to your mama. But this time, let it be you who creates those awe-inspiring dishes. Maybe you like to cook with a friend or partner, or perhaps you find cooking alone therapeutic. Whatever your style, dig deep and tackle a toughie—it's just time, and what better way to spend it than creating intense flavor?

"EARTHY LUSCIOUSNESS"

FIG + CHEESE + BACON

Sweet and delectable, figs become even tastier with the introduction of contrasting flavors. Cheese deepens, rounds out, and enriches the fig. Bacon adds the salty intensity needed to elevate all three flavors. Figs are best served with little or no cooking, leaving the most room for variation with the cheese component. Alternatives include a composed cheese course with pancetta and fresh figs or a terrine of cream cheese, fig jam, and crumbled bacon with crackers. These flavors also work well with poultry and game meats.

FIG

Fresh figs enjoy a short season from late summer throughout the fall months. They are also dried and are effectively rehydrated. Keep the liquid (usually water or wine) used to rehydrate dried figs for use in sauces and vinaigrettes or reduced to a syrup glaze. For fresh varieties and flavor versatility, look for the dark purple Black Mission or the light green Kadota.

CHEESE

In smaller proportions in a dish, cheese can serve as a flavor accent or to increase creaminess. Cheeses that taste great with figs include soft ones such as goat cheese, Brie, and blue cheese. These serve as a creamy, soft complement to the luscious fig. For aged hard cheeses such as Manchego or Parmesan, consider using a cooking technique to soften the bite—for example, make a cheese flan.

BACON

For this combination, bacon brings a savory, salty flavor, and sweetening techniques (such as brown sugar baking) would interfere with the intended impact. Save the bacon fat rendered from cooking and use to sauté and enrich starch dishes such as mashed potatoes.

"EARTHY LUSCIOUSNESS" RECIPE

Roasted Figs with Crispy Bacon and Goat Cheese Foam

This appetizer is a play on contrasting effects—warm and sweet fig with airy goat cheese foam and a crunchy bite of bacon. You'll need an iSi Stainless Steel Cream Whipper, available at gourmet food shops and online. This is a fun tool for many applications beyond making whipped cream, such as this savory foam.

Special equipment required

¼ cup (60 ml) cold white wine
   (such as Pinot Gris or dry Riesling)
¼ sheet gelatin (or ¼ teaspoon
   powdered gelatin)
1 cup (235 ml) heavy cream
½ cup (120 ml) milk
4 ounces (113 g) **soft goat cheese**,
   crumbled
Salt and pepper
4 slices **bacon**, cut in ½-inch
   (1¼ cm) pieces
8 fresh **figs** (Black Mission preferred),
   stems trimmed, sliced in half
2 tablespoons (40 g) honey

Place cold wine in a large bowl and add gelatin (sprinkle over, if using powder.) Set aside to soften.

In a saucepan over medium heat, heat cream, milk, goat cheese, salt, and pepper to taste. Cook, stirring until the goat cheese dissolves. Pour the hot cream mixture into the bowl with the gelatin; stir to combine. Push through a fine mesh strainer and cool. Pour into metal cream whipper and refrigerate until using.

Preheat oven to 300°F (150°C, or gas mark 2). Fry bacon in a skillet until crisp and drain on paper towels.

Place figs on a sheet pan and drizzle with honey. Roast for 5 minutes and then let cool slightly. Place roasted figs on a tray, put a few bacon pieces on the figs, and top with a dollop of goat cheese foam. Serve immediately, as the foam will quickly deflate.

Yield: 4 servings

 For a fun way to serve these, assemble them on individual spoons. Make them small enough to be one-biters and pass to your guests.

"BASKING IN THE SUN(CHOKE)"

SUNCHOKE + CHEESE + BITTER GREENS

Sunchokes' earthy, nutty taste is a unique flavor not to be missed. Combining these tubers with a rich, unctuous cheese and the flavor power of bitter greens makes for a wonderfully complex combination.

The application recipe prepares a warm sunchoke blini topped with broiled Bûcheron goat cheese and mizuna. Other options for this combination include preparing a gratin with sunchoke and Parmigiano-Reggiano cheese served with watercress or a creamy sunchoke soup with Brie crostini and arugula. These flavors pair nicely with fish, eggs, brussels sprouts, and mushrooms.

SUNCHOKE

Also known as Jerusalem artichokes, these vegetables are crisp and mild when raw and can be used as a substitute in recipes calling for jicama or water chestnuts. Because they are knobby and difficult to peel, they are best in rustic preparations or puréed, where peeling is not necessary.

CHEESE

Fresh goat cheeses, with their mild tang, work well with this combination, as do nutty flavors such as Parmigiano-Reggiano. For a smooth, creamy texture, try a soft cow's milk cheese such as Saint André. For a bold cheese flavor, try a blue cheese—Stilton or Gorgonzola would both provide a rich cheese entrance to the combination.

BITTER GREENS

Try endive, frisée, mizuna, arugula, radicchio, or cress (pepper or water). For maximum effect, use bitter greens raw, as an accent flavor. Or braise endive or radicchio to soften the flavor and texture and then build a dish around it.

Warm Sunchoke Blini Salad

This is a delicious salad course that you could also serve as a main vegetarian course. The blini batter requires some precision with measurement—that's why you must weigh the ingredients for this recipe.

Blinis

12 ounces (340 g) Yukon Gold potatoes,
    peeled and rough cut

14 ounces (400 g) **sunchokes**, rough cut

3½ ounces (100 g) all-purpose flour

2 teaspoons (8 g) Dijon mustard

⅓ cup (80 ml) heavy cream

1 teaspoon (1 g) thyme leaves

4 eggs, yolks and whites separated

Salt and pepper

2 tablespoons (28 g) butter

Toppings

2 tablespoons (28 g) unsalted butter

8 ounces (225 g) **sunchokes**, sliced

1 teaspoon (1 g) thyme leaves

Vinaigrette

½ shallot, minced

1 tablespoon (11 g) Dijon mustard

Juice of ½ lemon

3 tablespoons (45 ml) Champagne vinegar

2 tablespoons (30 ml) olive oil

Salt and pepper

To Serve

8 ounces (225 g) **goat cheese**

2 cups (110 g) **mizuna**

To prepare blinis: Cook the potatoes and sunchokes in salted, boiling water until soft. Drain well and pass through a food mill into a large bowl. Let cool. Add flour, mustard, cream, thyme leaves, and lightly beaten egg yolks. Whip egg whites to soft peaks and fold into batter. Season with salt and pepper.

In a skillet over medium heat, melt butter. Spoon 2 spoonfuls of batter in the skillet for each blini. Cook until golden on each side, 2 to 3 minutes per side. Repeat with all batter (it does not keep well for future use). Place on a sheet pan while preparing toppings.

To prepare toppings: In a skillet, melt 2 tablespoons (28 g) of butter over high heat. Add wedges of sunchokes and cook until golden. Add thyme leaves and cook for an additional 1 minute. Remove from heat and set aside.

To prepare vinaigrette: Whisk together the shallot, mustard, lemon juice, vinegar, olive oil, salt, and pepper to taste. Set aside.

To serve: Preheat broiler to high. Crumble the goat cheese over the blinis. Broil until browned and melting. Toss mizuna with prepared vinaigrette (you will have extra vinaigrette, which keeps well in the refrigerator). Place a mound of mizuna on each serving plate and top with blinis. Scatter cooked sunchokes on top of blinis and around greens. Serve immediately. Or you may prepare the blinis in advance and rewarm then in the oven.

Yield: 4 first-course portions

"SWEET AND PUNCHY"

PARSNIP + GINGER + BASIL

Sweet and mild parsnips with the fragrance and punch of ginger make a wonderful canvas for the addition of basil, lending an herbal, light clove and anise taste. Parsnips especially offer versatility—grate them into salads with a ginger-basil vinaigrette, roast them for side dishes, or purée them for a mashed potato–like consistency or for smooth soups. These flavors pair nicely with seafood such as scallops and white fish such as sea bass.

PARSNIP

Although it looks like a white carrot, a parsnip has a milder, sweeter flavor than its orange cousin. Use smaller, more delicate parsnips grated in salads. Otherwise, parsnips are well suited to cooking techniques—broiling, roasting, and braising to soften and increase the sugars. Look for smaller parsnips, as the larger ones can have a woody, bitter center you must remove.

GINGER

Fresh root is the way to go for this combination. To add the flavor of ginger without the texture or to evenly distribute in sauces, use ginger juice. Grate fresh ginger, wrap it in cheesecloth, and squeeze it over a small bowl. If you don't frequently use fresh ginger, you can freeze what's leftover. This actually makes the grating easier, too. Instead of cheesecloth, you can also try a garlic press.

BASIL

Dried basil loses most of its flavor, so stick with fresh whenever possible. When making basil oil, as in this recipe, use a small amount of fresh spinach leaves with the basil. You won't be able to taste the spinach, but it will help create a rich green color that offers a great contrast to the creamy white parsnips. Look for Thai basil as a variation and to accent the ginger for an Asian slant.

"SWEET AND PUNCHY" RECIPE

Sweet Parsnip Soup with Thai Basil Oil

This assembly takes your meal in an Asian direction, with the use of miso, mirin, and Thai basil.

Soup

½ cup (120 ml) mirin (cooking sake)

½ cup (120 ml) sweet white wine
    (such as Riesling)

1 cup (235 ml) water

2 tablespoons (32 g) miso paste

1 tablespoon (6 g) minced **fresh ginger**

1 tablespoon (10 g) minced fresh garlic

6 medium **parsnips**, peeled and chopped

2 small Thai chile peppers, split and
    seeds removed

1 can (16 fluid ounces, or 475 ml)
    unsweetened coconut milk

1 tablespoon (15 ml) rice vinegar

1 teaspoon (5 ml) fresh lemon juice

Oil

1 cup (24 g) **Thai basil leaves**

2 tablespoons (4 g) very green spinach
    leaves or parsley (optional, for color)

½ cup (120 ml) canola or grapeseed oil

Salt

To prepare soup: In a large saucepan, combine mirin, wine, water, and miso over medium heat. Whisk to dissolve the miso and bring to a simmer.

While miso broth simmers, sauté ginger and garlic in 1 tablespoon (15 ml) of oil until softened. Add to miso broth, along with parsnips and chiles. Cook over medium heat until parsnips are very soft. Remove from heat and remove chiles. Let cool slightly and purée in a blender with the coconut milk. Return to saucepan and stir in vinegar and lemon juice.

To prepare oil: In a saucepan of boiling water, blanch basil and spinach (if using) for 30 seconds. Immediately plunge into an ice bath. Remove after 1 minute and squeeze hard to extract as much water as possible from the basil (don't worry if it doesn't look like much; the quantity reduces drastically). Add to a blender with the oil and run for 2 minutes. Strain through a fine mesh strainer, add a pinch of salt, and use a funnel to pour into a small squeeze bottle. Oil will keep for up to 1 week in the refrigerator, but the color and flavor will diminish over time.

To serve soup: Reheat soup and ladle into serving bowls. Drizzle with Thai basil oil, unleashing your inner artist for an aesthetically pleasing presentation.

Yield: 4 servings

Make this dish more substantial by adding seared scallops and crusty bread.

"SMOKIN' CRUSTACEAN"

LOBSTER + CREAM + SMOKED PAPRIKA

The delicacy of lobster offers a rich taste that goes well served with a rich accompaniment—cream (though it's often butter). The cream smoothes out the lobster flavor without obliterating it. Smoked paprika has a deep, intense flavor that makes this combination both complex and delicious. For a twist on the lobster bisque, serve it in coffee cups, top with frothed milk (use an immersion blender), and call it "lobster cappuccino" that's sipped from a cup.

LOBSTER

Most lobster meat comes from the crustacean's claws and tail and can be baked, broiled, or fried. You can also cook the shell and body to make a flavorful stock, or use the red roe as a seasoning. This application needs the whole lobster; for a more straightforward approach, use the meat only to garnish a creamy potato-paprika stew.

CREAM

Use cream as the base for a savory sauce, soup, or starch. To decrease its proportion in the following bisque, add a starch such as potato—a twist on the classic brandade with lobster and smoked paprika.

SMOKED PAPRIKA

Paprika can range from slightly sweet to quite spicy. Smoked paprika has a distinct flavor key in chorizo and Spanish cooking and will add a layer of interest to the cream's monochromatic flavor.

Deconstructed Lobster Bisque

Pull out this rich, flavorful soup for special occasions.

2 **whole lobsters** (1½ pounds [680 g]
    each) or 3 pounds (1.4 kg) shrimp,
    unpeeled

2 cups (475 ml) water, plus more to stop
    lobster cooking

2 tablespoons (28 g) unsalted butter,
    plus more for sautéing meat

2 stalks celery, rough cut

2 large carrots, rough cut

1 whole yellow onion, rough cut

4 cloves garlic, rough cut, plus more for
    sauteing meat

3 ounces (85 g) tomato purée

½ bottle (375 ml) dry white wine

2 sprigs thyme

4 bay leaves

3 tablespoons (42 g) unsalted butter

3 tablespoons (23 g) all-purpose flour

¼ cup (60 ml) brandy

½ cup (120 ml) cream sherry

2 cups (475 ml) **heavy cream**

1 teaspoon (2.5 g) **smoked paprika**

Salt and pepper

Blanch lobsters in salted, boiling water for 12 minutes and then immediately plunge into an ice bath. Crack lobster shells and remove meat. Set meat aside. If using shrimp, peel shrimp and set meat aside.

Heat rondeau (a large pot, wide and low, with 2 handles) or other large pot to medium and add butter. When the butter stops bubbling, add lobster or shrimp shells, celery, carrots, onion, garlic, and tomato purée. Cook until lightly caramelized. Deglaze with white wine and add thyme, bay leaves, and water. Cook for at least 30 minutes. Strain stock and discard solids.

Melt butter in a skillet over medium-low heat. Stir in the flour and cook over low heat for 5 minutes, stirring frequently. Add to strained liquid and stir to distribute, allowing mixture to thicken.

Add brandy, sherry, and cream. Season with smoked paprika, salt, and pepper.

Sauté lobster or shrimp meat in additional butter, garlic, salt, and pepper. Serve with bisque.

Yield: 6 to 8 servings

Lobster is a luxury ingredient available fresh in many supermarkets. Don't be intimidated by buying a whole lobster! Get your money's worth by saving the shells, freezing if not using right away, and making a stock for this impressive combination.

"ALL WARM AND TOASTY"

POPPY SEED + SESAME SEED + ONION SEED

In this combination, fragrant and flavorful seeds create an intense spice blend, capturing what is great about an everything bagel—nuttiness, crunch, and spice.

The application recipe prepares a tasty flatbread. For other opportunities to use the combination, try it as a topping for homemade crackers, add it to creamy dips, or use it to season marinades or dressings. It is well suited to baking with breads, crackers, or crisps.

POPPY SEED

The nutty aroma and flavor of poppy seeds make these excellent for both savory and sweet applications. For savory uses, combine with noodles, fish, or vegetables. Also use in muffins, breads, or pastries. Those exposed to drug testing know poppy seeds; their consumption alone purportedly causes a positive result!

SESAME SEED

Use either white or black sesame seeds, which are similar in flavor, in this combination. Combining them provides visual interest, as the other seeds in this application are both black. Look to sesame oil to impart a rich sesame flavor when using the combination for marinades or vinaigrettes—just be sure to use a small amount, as it has a potent flavor.

ONION SEED

They are not actually related to the onion, but these small black seeds do have a slightly pungent, peppery, bitter flavor. They also are called nigella or kalonji seeds— helpful to know when shopping in an ethic market.

Fantastic Flatbread

This is not a difficult recipe, especially if you have a standing mixer with a hook attachment. Use it as the base for a pizza or sandwich or dip it into tzatziki.

¾ tablespoon (9 g) active dry yeast

1½ teaspoons (6 g) sugar

1 cup (235 ml) warm (not hot) water

3 cups (375 g) all-purpose flour, plus
   more if kneading

1 teaspoon (3 g) **poppy seeds**

1 teaspoon (3 g) **black sesame seeds**

1 teaspoon (3 g) **onion seeds**

1 tablespoon (18 g) salt (preferably
   sea salt)

1 tablespoon (15 ml) olive oil, plus more
   for cooking

Prepare the dough in advance by rolling out the balls and keeping them in the refrigerator overnight on a sheet pan. Or cook them in advance and reheat in the oven before serving.

Combine the yeast, sugar, and warm water in a small bowl. Put in a warm spot and let sit for 10 minutes, until foaming.

Meanwhile, in the bowl of a standing mixer, add the flour, poppy seeds, sesame seeds, onion seeds, and salt. Whisk to combine. Add the olive oil to the yeast mixture and then add to the flour mixture. Attach to the mixer and mix on low with the hook attachment for 10 minutes until smooth and elastic. Alternatively, knead by hand on a floured surface.

Add the dough to a buttered or oiled bowl, turn dough so the oiled side faces up, and cover with plastic wrap. Place in a warm environment until the dough doubles in size, between 1 and 1½ hours. Punch down and turn out onto a lightly floured surface.

Divide the dough into 8 even pieces; roll into balls. Roll each ball into ¼-inch (⅔ cm) circles. Heat a griddle or grill pan over medium-high heat. Brush flatbread with olive oil and cook for 1 minute, oil-side down, until bubbles rise to the surface. Turn and cook for an additional 1 minute and repeat with remaining dough. Serve warm.

Yield: 8 pieces

"SALTY RIPENESS"

SALT + FENNEL + MELON

Here we take three ingredients, apply each in a different way and then bring them back together for a finished dish. Salt (plus time, about two weeks) cures a duck breast into duck prosciutto. Fennel's frond becomes oil, its bulb shaved for a crunchy salad, and its pollen used to season the dish. The melon remains in its pure ripe simplicity, providing a sweet, melting texture. For a less complex way to use the combination, purée cooked fennel with ripe honeydew for a chilled soup sprinkled with pink Himalayan salt. These flavors are also nice with chicken, bitter greens, and figs.

SALT

For curing meat, salt inhibits the growth of bacteria. Use a coarse kosher salt for this, as you can more easily rub it off. For an all-purpose salt, try a finely ground sea salt. Or for a visually stunning effect, as in garnish for a soup or salad, pick up some pink-hued Himalayan or black volcanic salt.

FENNEL

The application recipe uses the bulb, frond, and pollen (save the stalks to make stock or cook with roast meats). The fennel bulb is the base of the plant and has a tough outer layer, which should be removed. The fronds are the frilly end of the fennel stalk and look like fresh dill. Fennel pollen is a specialty spice ingredient harvested from wild fennel. It appears like fluffy sand and has an intensely sweet-pungent flavor. It is expensive, but a little goes a long way, and its flavor can tweak an entire dish. Find it at gourmet or specialty food stores and online.

MELON

Look to honeydew, cantaloupe, or Crenshaw melons (all in the muskmelon family)—for their full flavors and smooth flesh. Ripeness, critical when using melon, can often be difficult to discern. Here are a few clues: Smell the melon for a sweet aroma. Touch the skin; it should feel smooth and slippery as the fruit matures (not hairy). Or ask someone who works in the produce section or at the farmers' market to choose one for you.

"SALTY RIPENESS" RECIPE

Duck Prosciutto Salad

The duck needs at least two weeks to hang in the refrigerator, so make sure to start this recipe early. Make the fennel oil the day before serving, so that plating this salad will take no time at all.

Duck Prosciutto

1 teaspoon (1.7 g) coriander seeds

1 teaspoon (2 g) whole black peppercorns

½ teaspoon dried thyme

½ teaspoon cumin seeds

1 bay leaf

¼ cinnamon stick

1 tablespoon (19 g) **kosher salt**

1 teaspoon (5 g) packed brown sugar

1 large duck breast (such as Moulard), trimmed of excess skin

Fennel Oil

Frond from 1 **fennel bulb**

¼ cup (7.5 g for spinach, 15 g for parsley) spinach or parsley leaves

⅓ cup (80 ml) canola oil

3 tablespoons (45 ml) almond oil (optional; use all canola oil if not using almond oil)

Salt

Salad

1 ripe **melon** (cantaloupe, honeydew, or both), very thinly sliced

1 bulb fennel, very thinly sliced (preferably on a Japanese mandoline)

Duck prosciutto, very thinly sliced

Large pinch fennel pollen

To prepare duck prosciutto: In a skillet, combine coriander, peppercorns, thyme, cumin, bay leaf, and cinnamon. Toast over medium-high heat until the mixture begins to smoke, 1 to 2 minutes. Allow to cool and then finely grind in a spice grinder. Mix with salt and brown sugar.

Rub duck breast with spice mix. Wrap tightly with plastic wrap and refrigerate for 24 hours. After 24 hours, brush off excess spices left on the breast and wrap the duck in cheesecloth. Leave enough extra cheesecloth to poke a string through; with the string, hang the duck, breast up, in the refrigerator for 2 weeks.

After 2 weeks, remove the cheesecloth and you have prosciutto. To slice very thin, freeze the prosciutto and use a meat slicer or very sharp knife.

To prepare fennel oil: In a saucepan of boiling salted water, blanch the fennel frond and spinach or parsley for 45 seconds. Use a small strainer to immediately plunge into ice water. Remove, squeeze any excess water, and add to a blender. Blend on high with the canola oil and almond oil (if using) and a pinch of salt for 1 to 2 minutes until thoroughly blended and no large green pieces remain. Push through a fine mesh strainer and use a funnel to fill a small squeeze bottle. Keep refrigerated until using.

To prepare salad: Spread out thinly sliced melon on a serving platter (or individual serving plates), covering the surface. Scatter with shaved fennel and drape slices of prosciutto over the top. Drizzle fennel oil over the entire dish and sprinkle with fennel pollen. Serve immediately.

Yield: 4 servings

"A SPANISH-ITALIAN RENDEZVOUS"

EGG + CHILE PEPPER + PROSCIUTTO

In this seemingly simple take on spicy eggs with cured meat, different cooking techniques create a multidimensional dish. In our application recipe, the soft-cooked poached egg lends its silky yolk and rich substance, a spicy piquillo pepper coulis adds a smoky burst of flavor, and the salty prosciutto wraps around a creamy salt-cod brandade (a pounded combination of fish, olive oil, garlic, milk and cream). The result is a satisfying, exciting play on textures and flavors. For a more straightforward preparation, try a frittata or egg salad sandwich using the same combination.

EGG

Using eggs in a breakfast application makes sense in this combination—scrambled, in omelets, or over-easy. If poaching, as in the following recipe, do so up to two hours ahead; just cook for slightly less time than for what the recipe calls. Trim the whites, and keep at room temperature on a sheet pan. Before serving, dot with butter, salt, and pepper and cook in an oven heated to 300°F (150°C, or gas mark 2) for about two minutes to warm through.

CHILE PEPPER

Use piquillo peppers, a Spanish variety, for their smoky, sweet piquancy. Don't go overboard with the heat here, so that someone eating this will be able to identify the other flavors. For a quick piquillo peppers substitute, try roasted red bell peppers with smoked paprika or Sriracha sauce.

PROSCIUTTO

This dry-cured Italian ham has an intensely concentrated flavor. Look for the prized prosciutto di Parma or check out Spanish varieties of cured hams such as Serrano or Iberian—the best of which feature pigs fed acorn-only diets.

"A SPANISH-ITALIAN RENDEZVOUS" RECIPE

Prosciutto-Wrapped Brandade with Chiles and Molten Egg

You need to start this the day before you want to cook it to properly prepare the salt cod. Although this dish has several different components, prep them a couple of hours in advance, giving you time to be with your dinner party guests (and not standing over a hot stove).

Special equipment required

Brandade

1 pound (455 g) salt cod

2 cups (475 ml) milk

2 cloves garlic, peeled

1 sprig thyme

Freshly grated pepper

1 bay leaf

2½ cups (595 ml) heavy cream, divided

½ cup (120 ml) dry white wine (such as Chardonnay)

2 pounds (900 g) small red potatoes, peeled

½ cup (112 g) unsalted butter

¼ cup (60 ml) olive oil

4 slices **prosciutto**, cut long (6 to 8 inches [15 to 20 cm]) and thin

To prepare brandade: Rinse the salt cod twice with cold water. Place in a bowl and cover with milk; refrigerate overnight.

The next day, preheat the oven to 350°F (180°C, or gas mark 4). Drain off milk and place salt cod in a roasting pan. Add garlic, thyme, pepper, bay leaf, 1½ cups cream (355 ml), and wine. Cook for 20 minutes or until fish flakes.

Meanwhile, cook potatoes in salted, boiling water. When soft, drain well (additional water is the enemy) and run through the food mill or ricer into a bowl. In a small saucepan, warm remaining 1 cup (235 ml) cream with butter and olive oil until butter melts. Slowly add warm liquid to the potatoes, stirring to incorporate, to ensure that the potatoes remain thick.

Once the fish comes out of the oven, remove thyme and bay leaf. Put fish and a bit of the poaching broth into the food processor and grind finely. With processor running, add remaining poaching liquid slowly, in a steady stream. Scrape fish mixture into bowl with potato purée and fold to combine. Let brandade cool.

Lay prosciutto slices out on a sheet pan. Place an oval scoop of the brandade on top of each slice, toward one end and then roll up the prosciutto. Prepare these up to 2 hours in advance but be sure to wrap them tightly with plastic wrap in the refrigerator to ensure that the prosciutto does not dry out.

(continued on next page)

Pepper Coulis

1 can (12 ounces, or 340 g)
   piquillo peppers
4 cloves garlic, peeled
1 tablespoon (15 ml) sherry vinegar
1 tablespoon (15 ml) olive oil
1 tablespoon (14 g) unsalted butter
Salt

Poached Eggs

4 **eggs**
2 tablespoons (30 ml) white vinegar
Salt

Add peppery greens such as arugula to give this dish a crunchy, contrasting flavor accent. Another option: Add garlic-rubbed crostini for a bit of crunch.

To prepare the pepper coulis: Preheat the oven to 350°F (180°C, or gas mark 4). Place peppers, with oil and juice from the can, in a roasting pan. Add garlic to pan and roast for 20 minutes. Scrape peppers and all liquid into a blender; add vinegar, oil, butter, and a pinch of salt. Blend until super-smooth. Adjust seasoning with additional salt, if necessary.

To poach the eggs: Fill a large saucepan (at least 3-inches [8 cm] deep with room for the boiling water to move) with water, vinegar, and a pinch of salt. Bring to a rolling boil. Break eggs into a small bowl—do not stir, mix, or otherwise disturb the eggs. With a slotted spoon, swirl the boiling water around the pan. As the water moves, slowly slide the eggs into the water, one at a time, while gently swirling the pan. This prevents the eggs from sticking to each other or to the pan's bottom. Watch closely and after 2 minutes use the slotted spoon to pull an egg out of the water. Gently touch it—it should feel soft in the center but otherwise set. If it doesn't feel ready, return it to the water and check again in 30 seconds. Remove eggs to a plate. Trim any dangling egg white with scissors.

To serve: Warm the brandade roll in an oven heated to 350°F (180°C, or gas mark 4) for 5 minutes, if it has been refrigerated. Warm pepper coulis and place in a small pool on a plate or shallow bowl. Top with brandade roll and perch poached egg on the very top. Carry carefully. Eat immediately.

Yield: 4 servings

"ONE RAD TOMATO"

TOMATO + FENNEL + MUSTARD

The versatile sweet tomato makes a great base for anise-flavored fennel and tangy mustard. Plus, you can employ many different cooking techniques to both the tomato and fennel. Use them both raw for a fresh salad with a Dijon fennel seed vinaigrette. Roast the tomato and fennel and season them with mustard seed for a brothy soup. Purée all three ingredients for a creamy soup. Or go the route of the application recipe and season a braised lamb dish with this ingredient combo. These flavors are well suited to fish, chicken, beef, or lamb.

TOMATO

Use ripe summer tomatoes raw, sauté for sauces or soups, or oven-dry them to increase their sugars and flavor intensity (an especially helpful technique when using off-season tomatoes). During winter months, opt for high-quality canned products. Also, look to tomato paste as a quick way to insert intense tomato flavor into braised vegetables and sauces.

FENNEL

How you handle the fennel will determine its impact in this combination. Fennel's most intense flavor comes when it is raw and crunchy. Sauté and simmer it to soften or roast it until it practically melts. Candy it for sweetness. Fennel seeds often flavor dishes and get used in salad dressings and rubs, with its frond used as an herb. Increase the fennel factor for savory dishes.

MUSTARD

For this combination, use either prepared mustard or mustard seeds to introduce the flavor. Dijon ranges from mild to spicy and adds a refined appearance to sauces and dressings. Whole grain introduces an additional textural element and has a more rustic flavor for meat applications. Use toasted mustard seeds for a light flavor in vinaigrettes or pickled fennel and tomatoes.

"ONE RAD TOMATO" RECIPE

Tomato-and-Fennel Braised Lamb Shanks

This dish needs several hours in the oven, so start early if you plan to eat it the same day you cook it. Enjoy the rich aroma that wafts from the kitchen as the meat cooks.

4 ½ pounds (2 kg) lamb shanks, trimmed

Salt and pepper

½ cup (65 g) all-purpose flour

6 tablespoons (90 ml) olive oil, divided

2 tablespoons (28 g) unsalted butter

1 cup (235 ml) dry red wine (such as Cabernet Sauvignon)

¼ cup (44 g) **whole grain mustard**

2 yellow onions, sliced into ¼-inch (⅔ cm) wedges

1 large leek, sliced in half lengthwise and cut into ¼-inch (⅔ cm) pieces (don't use the dark green part)

4 cloves garlic, chopped

½ cup (120 ml) Pernod (or other anise liqueur)

2 cups (475 ml) chicken stock

1 can (28 ounces, or 820 g) peeled whole **plum tomatoes**, drained

2 tablespoons (32 g) **tomato paste**

1 tablespoon (2.5 g) thyme leaves

2 bay leaves

2 bulbs **fennel**, cut into ¼-inch (⅔ cm) slices

Preheat oven to 375°F (190°C, or gas mark 5). Season all sides of the lamb shanks with salt and pepper and then dredge in flour to coat. In a large skillet, heat ¼ cup (60 ml) of olive oil and the butter over medium-high heat. Sear the lamb until brown, turning to cook all sides. Transfer shanks to a roasting pan.

In the skillet, spoon out all but a couple spoonfuls of fat. Over medium heat, deglaze the pan with the red wine, stirring to remove the stuck scraps. Whisk in the mustard and cook for 2 minutes. Coat the lamb shanks with the red wine–mustard mixture and set aside.

In a large, clean skillet, heat the remaining 2 tablespoons (30 ml) of olive oil over medium heat. Add the onions and leek and cook until tender, 5 minutes. Add the garlic and cook for an additional 2 minutes.

Add the Pernod and increase the heat to medium-high, cooking for 2 minutes. Add the chicken stock, tomatoes, tomato paste, thyme, and bay leaves; stir to incorporate the tomato paste. Add the sliced fennel to the roasting pan with the lamb. Pour the skillet's contents over the lamb and fennel and cover with aluminum foil. Braise for 2½ hours. Skim any fat off the surface of the liquid. Serve braised shanks with vegetables and cooking broth.

Yield: 4 servings

Serve these shanks in a shallow bowl with something such as polenta to absorb the delicious braising liquid.

"A WARM TOUCH OF FALL"

APPLE + HONEY + ALMOND

The leaves start to turn, a chill blows in, and apples are in season. This delicious combination uses apples primarily in desserts by adding intensely sweet honey (applied in a number of different ways) and the mild crunch of almonds.

The following recipe is for challah bread. This yeast-based bread requires several hours to make but is worth the effort. For a less time-intensive experience with the combination, buy a high-quality loaf of challah or brioche, make French toast, and top with an apple-honey-almond butter. Or try a strudel or coffee cake using the combination ingredients, whose flavors closely associate with cinnamon and nutmeg.

APPLE

This recipe uses dried apples, which provide a rich, concentrated flavor without adding moisture to the dough. For a different effect, substitute thinly sliced tart apples, such as Granny Smith. Or make a homemade honey-spiked applesauce topped with candied almonds.

HONEY

Make a bolder honey statement by using a dark, flavorful variety such as wildflower honey. To use honey in a savory application with this combination, braise lamb with chopped apples and honey, add spices such as curry or garam masala, and garnish with chopped almonds.

ALMOND

Almonds have much potential: as a textural element toasted and chopped, ground into a flour for sweet breads or cakes, in an extract for additional flavoring, or as a paste to create fillings for strudels and pastries. To maximize the almond flavor, combine two or more of these options.

"A WARM TOUCH OF FALL" RECIPE

Challah Bread with a Twist

If you have any leftovers (doubtful), this bread makes an incredible French toast.

½ cup (112 g) unsalted butter, divided,
   plus additional for bowl and pan

3½ cups (438 g) flour, plus more
   for kneading

½ teaspoon ground cinnamon

⅛ teaspoon ground nutmeg

¾ cup (175 ml) warm water
   (100°F [38°C])

⅔ cup (230 g) **honey**, divided

2 eggs, plus 3 yolks

2 teaspoons (8 g) active dry yeast

2 teaspoons (12 g) salt

1 cup (86 g) **dried apples**, diced

½ cup (55 g) **sliced almonds**

🧑 People of all faiths
can make this bread,
traditionally served in Jewish
homes on Friday night. Try it
for Rosh Hashanah, a holiday
in the fall that celebrates
the Jewish New Year. Apples
are in season then, just the
reason to indulge in a sweet
treat.

Butter a large bowl and set aside. Melt ¼ cup (55 g) butter in the microwave or a small saucepan and then allow it to cool. In a large bowl, combine half of the melted butter with the flour, spices, water, ⅓ cup (115 g) honey, eggs and yolks, yeast, and salt. Mix into a dough and then turn out onto a floured surface and knead for 10 minutes until the dough is smooth. Place in the buttered bowl, spread 1 table-spoon (14 g) melted butter on top, and cover with plastic wrap. Place in a warm environment for 1 to 2 hours until the dough doubles in size.

Punch down the center of the dough and place on a floured work surface. Pat out into a large circle, add apples, and knead briefly to distribute the fruit throughout the dough. Return the dough to the bowl, spread remaining melted butter over the top, and cover with plastic wrap. Return the bowl to the warm environment and let rise for 1 hour until it again doubles in size.

Preheat oven to 375°F (190°C, or gas mark 5). Butter a 9-inch (23 cm) cake pan. Remove dough from bowl and roll into a long rope (about 20 inches [51 cm] long). Lay the dough in a coil shape in the buttered pan. Cover with buttered plastic wrap and allow to rise for 30 to 45 minutes until it doubles in size.

Heat the remaining honey and ¼ cup (55 g) butter together in a small saucepan, stirring to incorporate the butter. Brush over the challah dough and sprinkle with sliced almonds.

Bake for 40 minutes until challah rises, browns, and firms. Let cool in the pan on a wire rack for 20 minutes and then remove and allow to cool completely.

Yield: 1 loaf

"MAGICAL MIREPOIX"

ONION + CARROT + CELERY

These veggies, the big three of French cooking, make up the set of flavors called mire-poix, the traditional base of countless dishes. When cooked together, the vegetables' deep, earthy flavor gives the taste that sets French cooking apart from other world cuisines.

The application of this combination cooks the onion, carrot, and celery together as a bed for braised short ribs. But it also works wonderfully to flavor stocks, stews, soups, and sauces. It is the perfect weapon to pull out when you don't know what to cook for a sauce—start with mirepoix and let the contents of your pantry lead the way!

ONION

To use onion for mirepoix, always peel and cut it the same size as the carrot and celery. Uniformity of size is important to maintaining flavor balance, as well as for aesthetics of the finished dish. Use yellow or white varieties, but stay away from sweet (such as Walla Walla) or red onions.

CARROT

For this application you scrub, not peel, the carrots, especially if using organic carrots. In a classic mirepoix, carrots equal 25 percent of the total proportion of vegetables. Don't skip the carrot; if you don't have one around, consider it a good opportunity to get to know your neighbor by asking to borrow one. Or better yet, always have carrots around; they mean a healthy snack or fresh juice always only moments away.

CELERY

Always buy the whole celery plant, versus the hearts only. Save the leaves to add to salad greens for an interesting flavor dimension. Purchase and use for mirepoix firm, crisp celery. Pitch limp or brown celery. In a pinch, substitute fennel stems (save them when using the bulb only) in the same amount as the celery.

"MAGICAL MIREPOIX" RECIPE

Mirepoix-Braised Short Ribs

Braised meats are the ultimate do-ahead. Work on the meat happens at the front end, leaving the oven to do the rest. Start this the day before you intend to serve it. Instead of having friends over and grilling, save your time together for conversation, not tending the fire. This is a great meal with simple potatoes and salad.

5 pounds (2.3 kg) beef short ribs

Salt and pepper

1 **carrot**, rough cut

1 rib **celery**, rough cut

2 **onions**, rough cut

2 cloves garlic, peeled

1 fennel stalk, rough cut

2 tablespoons (32 g) tomato paste

1 bay leaf

2 cups (475 ml) dry red wine (such as
 Syrah or Cabernet Sauvignon)

2 cups (475 ml) light chicken
 (or veal) stock

2 sprigs fresh thyme

½ cup (120 ml) Pernod liqueur (optional)

¼ cup (44 g) Dijon mustard

The ingredients of mirepoix don't depend on season, giving you easy access to incorporate the flavors into your everyday cooking. This short rib recipe is a rich, delicious meat perfect for the cold winter months, when it's a blessing to leave your oven on for hours.

Preheat oven to 275°F (140°C, or gas mark 1). Season short ribs liberally with salt and pepper. Heat a large skillet over medium-high heat, and brown meat on all sides. Remove meat and add to a roasting pan.

Using the same skillet used to brown the meat, lower the heat to medium and add carrot, celery, onion, garlic, fennel, and tomato paste. With a wooden spoon, distribute the tomato paste among the vegetables and continue to stir until the vegetables begin to caramelize. Remove from heat. Add bay leaf, wine, stock, thyme, and Pernod (if using).

Spread mustard over the top of the meat. Pour contents of skillet over the meat and into the roasting pan. Cover pan with parchment paper, then with foil. Cook for 2½ hours covered. Uncover and then cook for an additional 1½ hours. Allow to cool and then refrigerate overnight, uncovered.

The next day, preheat the oven to 350°F (180°C, or gas mark 4). Skim the fat from the top of the meat. Warm in the oven for 15 minutes or until the meat pulls easily out of the juices. Trim the fat off the short ribs and pull to remove the bones. Cut into desired portion sizes.

Strain the cooking broth into a saucepan, discarding the remains. Cook the broth over medium heat until reduced by two-thirds. Use for a sauce with the short ribs.

Yield: 4 servings

"A JUICY BITE"

BLUE CHEESE + PEAR + NUT

This combination certainly satisfies as a simple cheese plate with each ingredient in its purest, uncooked state. However, the classic combination of cheese, fruit, and nut transforms once the cooking technique for each component changes. The juicy, ripe pear softens the tang of blue cheese, and a salty nut provides a crunch that keeps the combination from becoming too soft. These flavors add nice flavor to pork preparations.

BLUE CHEESE

Most famous as Gorgonzola in Italy and Roquefort in France, blue cheese comes in various consistencies—from quite firm to soft ripened. Not typically suitable for melting, it is often served with bread or on salads, perfect with sliced pear and toasted nuts.

PEAR

Ripeness is critical when eating pears raw. This fruit poaches nicely, but handle with care any cooking methods you try on pears because of their high water content. Pear purées make a nice chilled soup or, as in this application, a sauce for a cheese soufflé. Select Bartlett pears, among the juiciest, for puréeing; choose a Bosc or Anjou if serving raw or cooking softly.

NUT

Toasting and roasting nuts intensifies their flavors. Nuts that accompany this combination nicely include almonds, pistachios, pecans, and walnuts—classic for their slightly bitter, complicated flavor.

"A JUICY BITE" RECIPE

Gorgonzola Soufflé with Pear-Walnut Sauce

When making a soufflé, it is important that the eggs come to room temperature. Before you even gather your other ingredients, pull five eggs from the refrigerator so that they will be ready to go when you need them.

Pear-Walnut Sauce

1 cup (100 g) **unsalted walnut halves**

2 cups (475 ml) heavy cream

Salt

½ vanilla bean, split lengthwise and
    seeds scraped

1 **pear**, peeled, cored, and quartered

1 cup (235 ml) white wine (such as
    Riesling or Champagne)

1 teaspoon (5 ml) vinegar (Champagne
    or apple cider)

Juice of ½ lemon

¼ cup (50 g) sugar

¼ cup (60 ml) water

Blue Cheese Soufflé

¼ cup (25 g) grated Parmesan cheese,
    plus extra for sprinkling

3 tablespoons (42 g) unsalted butter,
    plus extra to grease the dish

3 tablespoons (23 g) all-purpose flour

1 cup (235 ml) milk

4 egg yolks, at room temperature

3 ounces (85 g) **blue cheese** (such as
Roquefort or Stilton)

Salt and pepper

5 egg whites, at room temperature

Pinch cream of tartar

To prepare pear-walnut sauce: Combine walnuts, cream, a pinch of salt, and vanilla pod and seeds in a saucepan and cook for 15 minutes over low heat until the liquid reduces by half. Strain mixture with a slotted spoon or through a fine mesh strainer into a bowl, pushing down to extract as much liquid as possible. Discard walnuts and pod.

In a small saucepan, cook pear, wine, vinegar, lemon juice, sugar, and water over low heat until the pear is very soft. When the liquid reduces by 75 percent and becomes thick and syrupy, remove from heat. Transfer mixture to a blender and blend until very smooth. Transfer sauce to a bowl and whisk in the walnut cream. Cover and set aside until it reaches room temperature and is ready to use.

To prepare soufflé: Grease an 8-inch (20 cm) soufflé dish and sprinkle with Parmesan cheese. In a saucepan over medium heat, melt the butter. Add the flour, mix to combine, and continue to stir and cook for 2 minutes. Add the milk, whisking to combine and cook, using a rubber spatula to scrape the bottom and sides occasionally. Cook until the mixture is thick and smooth. Remove from heat and whisk in the egg yolks one at a time. Stir in the cheese, season with salt and pepper, and set aside.

With the whisk attachment of a standing mixer, beat the egg whites and cream of tartar until thick, stiff, and glossy. Add about one-quarter of the stiff egg whites to the cheese base and fold in gently. Add the remaining whites and carefully fold until just incorporated. Fill prepared soufflé dish and bake for 30 minutes until soufflé browns and rises. Serve immediately with pear-walnut sauce.

Yield: 4 servings

"LIGHT, BRIGHT, AND SPARKLY"

LEMON + FENNEL + NUT

The anise-flavored fennel offsets the tang of lemon, which gets a depth of richness with nuts. Which nut variety you select affects this combination—the distinct taste of hazelnuts and pine nuts come through more prominently than macadamias or almonds. Also, Meyer lemons —much sweeter than regular lemons—yield more juice and produce a sweeter outcome. Finally, switching up the cooking techniques for the fennel can completely change this combination.

Here we assemble the ingredients for a dessert, but you could easily put them together as a light salad as well.

LEMON

Reductions and vinaigrettes often call for lemon juice; savory and dessert applications both utilize lemon segments. Zest contains concentrated flavor used in everything from cocktails to candy. Any of these lemon preparations could work effectively in this combination; try to layer the lemon flavor—in other words, use multiple forms of lemon—for the most impact.

FENNEL

Fennel's most intense flavor comes when it is raw and crunchy. Sauté and simmer it to soften or roast it until it practically melts. Candy it for sweetness. Fennel seeds often flavor dishes, with its frond used as an herb. Increase the fennel factor for savory dishes.

NUT

Toasting and roasting nuts intensifies their flavors. Eat them raw out of your hand, turn them into butters or brittle, or use them to add a textural note to a complex dish. For this pairing, think of pine nuts for a bold entrance or almonds for a less dominant flavor.

The play on textures in this recipe—creamy, icy, crunchy—enhances this combination of flavors and makes for a light dessert perfect after a heavy meal. You can prepare all of the components in advance, so it is a nice dessert to serve while entertaining guests.

"LIGHT, BRIGHT, AND SPARKLY" RECIPE

Lemon Pudding Parfait

You may prepare all pieces of this refined, grown-up dessert in advance. It is also great year-round, as its components are readily available during all seasons.

Lemon Pudding

2 tablespoons (16 g) cornstarch

½ cup (100 g) sugar

1 cup (235 ml) whole milk, divided

1 cup (235 ml) half-and-half

¼ cup (60 ml) **fresh lemon juice**

2 tablespoons (28 g) unsalted butter

Almond Ice Milk

2 cups (220 g) **sliced blanched almonds**

½ cup (100 g) sugar, divided

1 quart (945 ml) milk

Candied Fennel

1 **fennel bulb**, very thinly sliced

1½ cups (355 ml) simple syrup (page 183)

To prepare lemon pudding: Whisk together cornstarch and sugar in a medium-size saucepan. Add ½ cup (120 ml) milk to the saucepan and form into a smooth paste with a rubber spatula. Whisk in remaining milk and half-and-half and bring to a boil over medium heat, whisking constantly. Cook until thickened, using a rubber spatula occasionally to scrape the bottom and sides of the saucepan.

Remove from heat and whisk in lemon juice and butter. Scrape into a bowl and press plastic wrap directly onto the pudding's surface. Refrigerate for at least 4 hours and up to 1 day prior to serving.

To prepare almond ice milk: Place an 8-inch (20 cm) or similar-size baking pan in the freezer to chill. Preheat oven to 350°F (180°C, or gas mark 4). Spread almonds out on a sheet pan in a single layer. Toast almonds until dark brown, 8 to 10 minutes. Remove and cool completely.

Grind almonds with ¼ cup (50 g) sugar in a food processor until coarsely ground. Transfer nuts into a medium-size saucepan with remaining sugar and milk. Bring to a boil, stirring to dissolve the sugar. Remove from heat and allow to cool to room temperature.

Strain almond liquid and discard the solids. Remove chilled pan from the freezer and pour in almond milk. Return to freezer and set timer for 1 hour. Stir with fork or bench scraper after 1 hour and then reset your timer for another 1-hour interval and stir again. Continue to stir at 1-hour intervals until no liquid remains and ice milk has been chopped to a uniform size. This may take up to 4 hours, depending on your freezer's temperature. Wrap with plastic wrap until ready to serve. Prepare at least 1 day before serving or up to 1 week prior.

To prepare candied fennel: Preheat oven to 325°F (170°C, or gas mark 3). Fill a large saucepan with water and bring to a boil. Add sliced fennel and cook for 2 minutes, until just starting to become tender. Drain off water and spread fennel out in a single layer on a sheet pan. Cover with simple syrup.

Cover tray with a layer of parchment paper, then a layer of foil. Cook for 30 minutes, remove foil and parchment, and return to oven. Cook uncovered for another 30 to 60 minutes until the fennel is translucent and the simple syrup thickens. Cooking time will depend on how thinly the fennel is sliced. Allow to cool before serving. Keep covered in the refrigerator for up to 1 week until ready to serve.

To assemble: Scoop pudding into a fluted glass or ramekin, leaving room at the top for the other ingredients. Use the back of a spoon to level the surface of the pudding. Top the pudding with almond ice milk (2 to 4 tablespoons [30 to 60 ml], depending on your serving container). Top with candied fennel and serve.

Yield: 4 servings (1 quart [475 ml] ice milk)

"CHERRY CONTRASTS"

CHERRY + YOGURT + SESAME

In this combination of flavors, the ingredients play against one another for an outstanding richness of taste. Cherries bring a burst of sweet fruit to the tang of yogurt and the nutty depths of sesame.

The application recipe prepares a tart cherry soup with yogurt sorbet and sesame tuile (a thin, crisp, rounded cookie). Other options include preparing a sesame oil–spiked tart shell for a sweet yogurt-and-cherry-filled tart or baking the combination into a sweet bread and sprinkling with sesame seeds. These flavors pair well with honey, cinnamon, and ginger.

CHERRY

Try sour cherries (as opposed to the common sweet varieties) in this combination for their wonderful, tart flavor. These are also known as pie cherries, which are in season for just a sliver of time, about one month around July. Most end up in cans, but look for fresh sour cherries such as Montmorency, Morello, or Early Richmond at farmers' markets.

YOGURT

Use plain unsweetened or vanilla yogurt for this combination. Greek-style yogurt, increasingly available in regular grocery stores, is thicker and richer than regular yogurt because its whey is removed before packaging. Its consistency and flavor make it a great choice for cooking.

SESAME

Use the seeds to impart a nutty accent to dishes. White and black are similar in flavor, and combining them provides visual interest. Add sesame oil, which is quite strong, with a light hand. Use both the seeds and the oil to create a depth of sesame flavor in a dish, but in this combination, let sesame play a background role so that the other flavors stay in balance.

Chilled Cherry Soup with Frozen Yogurt and Crispy Cookies

Prepare all of these components a day in advance, making it easy to serve for guests. If you can't get fresh sour cherries, substitute sweet ones, such as Rainier.

Special equipment required

Soup

1 pound (455 g) **sour cherries,** pitted
⅔ cup (160 ml) water
½ cup (100 g) sugar
Pinch ground cinnamon

Frozen Yogurt

4 cups (920 g) **plain Greek-style yogurt**
1½ cups (355 ml) simple syrup
   (page 183)
Juice from ½ lemon

Sesame Tuiles

2 tablespoons (30 ml) milk
6 tablespoons (84 g) unsalted butter
⅓ cup (67 g) sugar
2 tablespoons (30 ml) light corn syrup
⅛ teaspoon sesame oil
1 cup (144 g) **white sesame seeds**
¼ cup (36 g) **black sesame seeds**

This is a summer dessert, when fresh cherries are in season. It is also great for hot weather because the tuiles require only a short amount of oven time.

To prepare soup: Place cherries, water, sugar, and cinnamon in a saucepan and bring to a simmer over medium heat, stirring to dissolve the sugar. Transfer to a bowl, cover with plastic wrap, and chill for 4 hours (or overnight).

To prepare frozen yogurt: Whisk all three ingredients together and then process in an ice cream machine according to manufacturer's directions. Store tightly covered in the freezer.

To prepare sesame tuiles: In a saucepan, bring the milk, butter, sugar, corn syrup, and sesame oil to a simmer. Stir in the sesame seeds. Transfer to a bowl, cover with plastic wrap, and refrigerate for 4 hours (or overnight).

Preheat oven to 375°F (190°C, or gas mark 5). Line a sheet pan with a nonstick Silpat liner or parchment paper. Scoop 1-inch (2.5 cm) balls of sesame batter (a small ice cream scoop works well for this) and place 2 inches (5 cm) apart on prepared tray. Bake until golden, 10 to 12 minutes. Allow to cool completely. Store in an airtight container for up to 2 days.

To serve: Divide soup among bowls (or a stemmed glass such as a martini glass, which makes a nice presentation) and top with a scoop of frozen yogurt and a sesame tuile. Serve immediately.

Yield: 4 servings

CHAPTER 4

Bright & Light

A bright balloon gently floats from your hand up into a beautiful blue sky dotted with white, fluffy clouds. The effect of these vibrant colors—a red balloon against an azure background—encapsulates what this chapter is about: bright and light. Do not expect this chapter to promote a diet or particular way of eating. It is about flavors, preserving and creating the intensity of food without added weight.

Food is typically weighed down by added fat—usually in the form of butter, cream, or animal fat. We often call food enriched in this way "heavy." This chapter focuses on pure flavors without reliance on enrichments. Using acid in preparing these recipes—generally achieved with citrus, vinegar, or naturally high-acid ingredients such as tomato— creates a light yet flavorful effect. Expect to find this theme throughout these recipes.

This chapter focuses on fruit, vegetables, and seafood—the building blocks of bright and light cooking. In their natural state, these foods contain little of the fats that can weigh down other dishes. We often team them up with fresh herbs such as mint, basil, and parsley or spices such as saffron, chiles, and vanilla, which contribute a depth of flavor while maintaining a bright, pure profile. These are not boring dishes—try the Chilled Soup with a Hint of Hot (page 127) made with cucumber, basil, and mint to see for yourself that light does not have to mean bland.

Handling ingredients lightly in terms of cooking technique, leaving them uncooked or lightly cooked, can contribute to a lighter flavor as well. Cooking an ingredient or finishing a dish frequently introduces fats, so the less you work with an ingredient, the more likely you'll end up with a lighter dish. These types of recipes typically work best in the hot-weather months because they don't require you to slave over a hot stove. Also at that time of year, they are most satisfying.

Don't assume that desserts aren't part of this chapter. Fruits such as strawberries, apricots, and oranges are natural meal-enders. Heighten their sweetness to satisfy any sweet tooth. One of the brightest, lightest recipes in this chapter could be Killer-Triple Berry Salad (page 123). It is exactly the type of flavor that could lift you off of the ground—just like that balloon let loose into the blue sky.

"LICORICE WITH SASS"

APPLE + FENNEL + LEMON

Pairing fruit with vegetables, as in this combination, often provides delicious contrast in flavor and texture. The tart, crisp apple joins with the crunchy, anise-like fennel, combining sweet and savory tastes. Lemon brings balance, providing the acid needed to make the flavors dynamic. The technique you choose will greatly affect the combination flavors, as both apple and fennel are versatile in preparation methods.

This application takes advantage of a simple, uncooked salad. But experiment with cooking the ingredients to change the profile—roasted apple, fennel, and preserved lemon work nicely for a side dish, for fish, or puréed into a soup. Or take this combination to the sweeter side and make lemon panna cotta with candied fennel and apple compote.

APPLE

For purity of flavor, extract apple juice in a vegetable juicer or seek out a good-quality natural apple juice. For the greatest texture and crisp flavor, opt for raw apples such as Pink Lady or Gala for sweetness or Granny Smith apples for tartness. Cooking softens the intensity of the apple's flavor and texture. To let the apple flavor dominate, make an apple compote accented by fennel frond and lemon rind.

FENNEL

Fennel's most intense flavor comes when it is raw and crunchy. When roasted, it transforms into a soft and mellow taste. Fennel seeds often flavor dishes, its frond used as an herb. Take advantage of the whole vegetable, even if the recipe does not call for it. The following application calls for the bulb only, but cut off the fronds to use as a flavorful herb you can add to the vinaigrette. Reserve the long stalks to add to other vegetables when making stocks and soups.

LEMON

Typically, in this combination, lemon serves as an accent component. The bright acidity of the lemon acts as a natural foil to the tart-sweet flavor of apples.

"LICORICE WITH SASS" RECIPE

Perfect-for-Fall Salad

Prepare this salad shortly before serving to prevent the apple and fennel from oxidizing and becoming discolored.

Vinaigrette
Juice of 1 lemon
Zest of 1 lemon
½ shallot, finely diced
¼ cup (60 ml) Champagne vinegar
1 tablespoon (13 g) sugar
¼ cup (60 ml) light, mild oil (such as grapeseed or canola)
Salt and pepper

Salad
2 **apples** (Pink Lady preferred), peeled
1 **fennel bulb**
1 bunch watercress, stems trimmed

To prepare vinaigrette: Mix together the lemon juice and zest, shallot, vinegar, and sugar in a small bowl. Whisk in oil and adjust seasoning with salt and pepper. Set aside.

To prepare salad: Keeping the apple whole, make thin slices until you reach the core. Turn apple to the other side and continue to make thin slices until you slice the entire apple. Using a Japanese mandoline helps to make thin, uniform slices. Make stacks of 4 to 5 slices and cut into matchstick-size batons. Slice fennel bulb in the same manner, thinly and then into matchsticks.

Place watercress in large bowl and top with apple and fennel. Sprinkle with salt and pepper and then drizzle with lemon vinaigrette and gently toss. Serve immediately.

Yield: 4 servings

Apples are in season during the fall months, just as fennel starts to show up. Serve this as a refreshing accent to lighten up the chill entering the air.

"EARTHY WITH A SIDE OF CRISP"

CUCUMBER + BEET + CABBAGE

The light, crisp cucumber meets other bright vegetables to create unique flavor play and crisp textures. Of the bunch, cabbage is the most versatile ingredient—leave it raw and crunchy in salads or cook it down for stews and side dishes. Most recipes use cucumbers raw, but beets need to be cooked.

Pickling all three ingredients is an option, but this application creates an amazing salad using raw cucumber and cabbage and roasted beets. These flavors work nicely with fish, especially tuna or light meat such as chicken.

CUCUMBER

Eat them raw, slice them up and serve with dip, or chop them and throw into salads. Because of cucumber's high water content, this vegetable generally doesn't cook effectively. For the best tasting cucumbers, wait for the summer locals to turn up at your farmers' market (or in your garden). To increase the impact of the cucumber, leave it raw and crunchy and serve with the other cooked and soft ingredients; it will provide a noticeable textural contrast.

BEET

Unless you use beets solely as a natural food coloring, cook them before eating. Beets keep well once cooked, so prepare them in advance and refrigerate them until use. They are also available precooked in the refrigerated section of your grocery store. To make this combination more about beets, make a borscht-style soup (a beet-based soup of Eastern European origin often served cold) topped with crunchy cucumber and cabbage.

CABBAGE

Cabbage is most flavorful when raw as in salads or a sandwich wrap. It is cooked to soften its flavor in soups and stews. All of these preparations work with this combination; just bump up the quantity to increase the cabbage's flavor against the strong beet.

"EARTHY WITH A SIDE OF CRISP" RECIPE

Pretty in Pink Salad

*The beets turn this slaw-like salad a charming pink color. You can make all of the components in advance,
but don't dress the salad until just before serving or the cabbage will lose its crispness.*

Tomato vinaigrette

1 clove garlic, peeled

1 small shallot, peeled and chopped

1 tablespoon (2.4 g) fresh thyme leaves

1 ripe Roma tomato, cut into quarters

¼ cup (60 ml) sherry vinegar

Salt and pepper

¼ cup (60 ml) extra-virgin olive oil

Salad

1 small **beet**, roasted, sliced, and cut
   into small cubes (or purchased already
   roasted)

2 tablespoons (30 ml) olive oil

Salt

½ **cucumber,** peeled and seeded

2 cups (180 g) thinly sliced
   Napa cabbage

1 large bunch watercress, stems trimmed,
   or arugula

Pepper

¼ cup (60 ml) tomato vinaigrette

To prepare the tomato vinaigrette: In the bowl of a food processor,
add the garlic, shallot, thyme, and tomatoes. Blend until chopped and
no chunks remain. Add the vinegar, salt, and pepper and blend. With
the motor running, add the olive oil until blended.

To prepare the salad: First roast the beet. Preheat the oven to 375°F
(190°C, or gas mark 5). Drizzle beet with olive oil and salt. Cook for
40 minutes or until tender. Peel beet before using.

In a large bowl, combine cucumber, beet, cabbage, and watercress.
Sprinkle with salt and pepper and toss with vinaigrette.
Serve immediately.

Yield: 4 servings

🔧 Change up this recipe
by using your favorite
vinaigrette to dress it. Just
be sure to use something
light in flavor that won't
compete with the other
components.

"INTENSE REFRESHMENT"

BERRY + CITRUS + CHEESE

Fruit often requires little preparation. In fact, lengthy preparations can sometimes take away from its pure flavor. In this combination, berries and citrus pair up with the contrast of cheese. It is a combination that maintains the sweet integrity of the fruit but also results in a gratifying, complex flavor.

The application recipe prepares a fruit-intensive green salad with raspberry vinaigrette dressing. For another option, take the combination in a dessert direction with cheesecake and fruit compote.

BERRY

Opt for the berry that looks the ripest and freshest. Before use, you'll need to stem and cut strawberries, but only wash blueberries, blackberries, or raspberries. Use frozen berries only when making a sauce; otherwise, stick with fresh.

CITRUS

Use oranges and grapefruit predominantly as a sweet accompaniment to the berries. Slip in a little lemon or lime for a tart accent. To use citrus segments in your cooking and take advantage of their juicy fruit only, "supreme" them. To do this, using a knife, cut off both ends and then the side peel of the citrus, removing all visible white pith with your knife. Then cut closely to the side of each segment, leaving pieces with no peel attached.

CHEESE

Blue cheese is ideal for this combination because it provides a rich, tangy flavor to contrast the sweetness of the fruit. Point Reyes in California makes a nice domestic blue cheese; the Original Blue is a versatile choice. For a milder-flavored cheese, look for a soft-rind goat cheese. Crumble and use it in the same way as a blue cheese.

"INTENSE REFRESHMENT" RECIPE

Killer Triple-Berry Salad

This salad has all the components of a complex salad—juicy, sweet fruit; salty, tangy cheese; flavorful greens; and crunchy, candied nuts. Make the dressing and segment the citrus ahead of time.

Raspberry Vinaigrette

1 shallot, chopped

2 cloves garlic, chopped

1 tablespoon (11 g), plus 1 teaspoon
   (4 g) Dijon mustard

1 tablespoon (13 g) sugar

1 tablespoon (20 g) honey

¼ cup (60 ml) raspberry vinegar

2 tablespoons (15 g) **fresh raspberries**

2 tablespoons (18 g) **fresh blackberries**

Juice from 1 lemon

¼ cup (60 ml) grapeseed oil

¼ cup (60 ml) olive oil

Salt and pepper

Salad

2 cups (110 g) chopped sweet greens
   (such as red oak, romaine, or butter)

2 cups (110 g) chopped spicy greens
   (such as frisée, watercress, or mizuna)

Salt and pepper

½ cup (75 g) **strawberries**, quartered

½ cup (65 g) **fresh raspberries**

½ cup (75 g) **fresh blackberries**

2 **oranges**, supremed (page 121)

1 **grapefruit**, supremed

½ cup (60 g) crumbled **blue cheese**

¼ red onion, thinly sliced

1 cup (100 g) candied walnuts

To prepare raspberry vinaigrette: Place the shallot, garlic, mustard, sugar, honey, vinegar, raspberries, blackberries, and lemon juice in a food processor. Run for 1 minute or until finely blended, scraping the sides twice or three times. With the processor running, add the oils slowly, in a steady stream. Adjust seasoning with salt and pepper. Keep covered in the refrigerator for up to several weeks, until ready to use.

To prepare salad: Place greens in a bowl and sprinkle lightly with salt and pepper. Evenly distribute the remaining ingredients on top of the greens, dress with raspberry vinaigrette, and serve immediately.

Yield: 4 servings

Though citrus peaks during the winter, it is available during the summer months as well, when berries taste their best. This is a flexible salad—use what looks best, and it will taste great.

"ROSY RED REFRESHMENT"

WATERMELON + TOMATO + MINT

These flavors may seem strange together, but they have an amazing ability to play off of each other for a bright, satisfying effect. The crisp, sweet watermelon and the ripe, juicy tomato get a flavor jolt from the fresh mint. It all comes together for an interesting summer mouthful. These ingredients benefit from raw uses, which preserve the crisp texture and pure flavor.

The application recipe prepares a light summer salad that is pure refreshment. Another option: Prepare a cool gazpacho (the cold, tomato-based soup) with the combination; just be sure to use more watermelon than tomato to ensure the watermelon's flavor shines. Other flavors that work well with the combination include balsamic vinegar and salty cheese, such as ricotta or feta.

WATERMELON

Seek out seedless varieties (which actually contain very small white seeds) for easier eating, especially if you plan to purée the fruit or cut it into specific shapes. At the height of the summer season, you can likely find different-colored varieties such as yellow and orange, which provide a fun presentation.

TOMATO

For this combination, try to get tomatoes in the same color as your watermelon—red, yellow, and orange. Part of the fun here comes from the surprise of not knowing exactly what you are biting into! Use a melon baller (or cut into same-size shapes) to contribute to the surprise. Store tomatoes at room temperature and never put them in the refrigerator—it compromises their texture.

MINT

Prolific in the home garden, fresh mint can quickly take over a yard. To use up yours consider making flavored simple syrup to add to iced tea, fruit salads, roasted fruit, or cool soups. Boil equal parts water and sugar, remove from heat, add fresh mint, and allow to cool in the syrup. Strain, cover, and refrigerate. It will last for months.

"ROSY RED REFRESHMENT" RECIPE

Multicolored Summer Salad

For this recipe, you can use a melon baller to scoop the watermelon. Try to get small cherry tomatoes to give the watermelon and tomato a similar appearance. Alternatively, dice them both in a similar size.

2 cups (310 g) **watermelon** balls, preferably multicolored

½ shallot, finely chopped

1 tablespoon (3 g) finely chopped chives

1½ cups (225 g) **cherry tomatoes**, preferably multicolored

⅓ cup (80 ml) rice vinegar (or white wine vinegar)

⅓ cup (32 g) chopped **fresh mint**

¼ cup (60 ml) olive oil

1½ teaspoons (6 g) sugar

Salt and pepper

4 cups (220 g) mâche greens (or chopped butter lettuce),
    or other grass

Prepare watermelon with the melon baller and keep separately in the refrigerator until ready to serve. Sprinkle shallot and chives over tomatoes and set aside. Whisk vinegar, mint, oil, and sugar in a small bowl. Season with salt and pepper.

To serve, gently combine watermelon and tomatoes. Place salad greens in a bowl, top with watermelon and tomato balls, drizzle with mint vinaigrette, and serve immediately.

Yield: 4 servings

This is a summer-only dish—don't even try it with winter offerings available in the grocery store. For additional flavor components, drizzle with aged balsamic vinegar or crumbled feta cheese.

"AN EVEN COOLER CUCUMBER"

CUCUMBER + BASIL + MINT

This combination of light summer flavors, when blended, offers a fresh and herbal flavor profile. Cucumbers are crisp, mild, and provide a lovely canvas for the summer brightness of basil and mint. These components work well in a salad, basil-mint vinaigrette over chopped cucumbers and greens, or as a soup, such as in the following recipe. Flavors that pair well with this combination include light fish such as halibut and rich grilled meats such as lamb kabobs, which offer a contrast.

CUCUMBER

The recipe prefers an English cucumber. What difference does it make? English cucumbers tend to be sold shrink-wrapped, so they lose less water and don't require waxing so you don't have to peel the wax off. English cucumbers also are considered semiseedless and purportedly result in fewer burps for the eater.

BASIL

Genovese basil, the most common variety found fresh in grocery stores, works great in this combination. If you can find lemon basil, it's certainly an exciting addition. Lemon basil has an awesome citrus aroma and pungent flavor that adds another level of intensity to the other ingredients.

MINT

As a rule, use fresh mint in cooking. At some of the world's most cutting-edge restaurants, chefs administer this rich aroma to diners in their dining rooms (as in a spritz over the table), creating a total sensory experience. One simple way to do this at home: Place fresh mint in vases on your table when serving dishes that feature this herb.

"AN EVEN COOLER CUCUMBER" RECIPE

Chilled Soup with a Hint of Hot

This cool soup comes together easily with only a blender. To cool quickly,
put soup in a bowl and place over an ice bath (a bowl of ice water).

1 cup (230 g) sour cream

½ cup (20 g) **fresh basil leaves**

½ cup (48 g) **fresh mint leaves**

¼ cup (40 g) chopped red onion

¼ cup (60 ml) water

½ cup (120 ml) semisweet wine (sparkling preferred,
    such as an effervescent Muscat)

1 tablespoon (15 ml) rice vinegar

1 large **cucumber** (preferably English), peeled, seeded, and chopped

1 tablespoon (15 ml) lemon juice

½ teaspoon Sriracha sauce

Salt

In a blender, purée the sour cream, basil, mint, red onion, water, wine,
and vinegar. Add the cucumber and blend until still slightly chunky.
Add lemon juice and Sriracha sauce to incorporate and then season
with salt. Chill thoroughly before serving.

Yield: 4 servings

This is a summer dish,
when all three featured
flavors are in season,
perhaps even growing in your
garden. In keeping with the
summer swelter, preparing
this flavorful soup requires
no cooking.

"ANCIENT AROMATICS"

ORANGE + SAFFRON + VANILLA

This delicately subtle combination has a rich bouquet that can create several different results. The orange's sweet acidity complements the intense orange color of the saffron and its hay-like, sometimes bitter flavor. The vanilla's aroma contributes to the bouquet, imparting a familiar, complex flavor. Use this combination as a braising liquid for fish or meat or for a sauce. Or take it to sweeter applications such as baked into a cake or poached with pears.

ORANGE

Using fresh-squeezed juice is always better—its flavor is brighter and more pure. Because recipes often call for orange zest in combination with the juice, you'll likely need an orange available anyway. Using a bottled juice won't ruin your dish; if you must go this route, use 100 percent juices that don't contain added sugar or other preservatives. To amplify the juice and zest flavor, use an orange oil or orange flower water in small amounts—they are very concentrated and require a delicate hand.

SAFFRON

Yes, saffron is expensive, but threads of saffron add exotic flavor to any dish, and a little goes a long way. In fact, be careful not to add too much to your cooking or you may end up with a medicinal-tasting dish. Always buy the threads, not powder, and store in the refrigerator for maximum shelf life. Add the threads to liquid to make the most of their flavor and impart a honey aroma and rich orange color. Stretch your investment by infusing your own olive oil to use in many dishes.

VANILLA

Do not substitute vanilla extract for vanilla beans in this combination. Look for Tahitian Gold or Bourbon vanilla beans, which are full of flavor. The seeds contain the flavor, so look for plump, juicy beans to add to the dish.

"ANCIENT AROMATICS" RECIPE

Braised Salmon

Light, mysterious, and full of flavor, this salmon preparation goes perfectly with savory mashed potatoes.

4 salmon fillets, skin on
   (6 to 8 ounces [170 to 225 g] each)

Salt and pepper

¼ cup (55 g) unsalted butter, divided

1 onion, finely chopped

1 bulb fennel, diced

2 tomatoes, cored and chopped

Zest of 1 **orange**, finely grated

½ cup (120 ml) **orange juice** (preferably
   fresh squeezed)

½ teaspoon **saffron threads**

1 **vanilla bean**, sliced in half lengthwise,
   seeds scraped

2 sprigs fresh thyme

1 cup (235 ml) water

1 cup (235 ml) light- or medium-bodied
   red wine (such as Pinot Noir)

Preheat oven to 350°F (180°C, or gas mark 4). Season both sides of salmon with salt and pepper. In a skillet over medium-high heat, melt 2 tablespoons (28 g) butter and sear salmon, skin-side down for 2 minutes until the skin crisps. Transfer salmon to a plate and set aside.

In the same skillet, add remaining butter and cook onion and fennel until translucent and soft. Add tomatoes, zest, orange juice, saffron, vanilla pods and seeds, thyme, water, and wine. Also, tilt into the skillet any juices from the plate with the salmon. Bring to a simmer and cook for 5 to 10 minutes until sauce reduces by two-thirds.

Remove from heat and place pieces of salmon on top of the reduction. Bake uncovered for 8 to 10 minutes or until the salmon is cooked through but still slightly translucent in the center. Serve immediately.

Yield: 4 servings

Use this recipe for other types of fish as well, such as sea bass. For a vegetarian option, substitute marinated tofu for the fish.

"WOODS IN SUMMER"
ASPARAGUS + ORANGE + OREGANO

The flavor of orange brightens and naturally sweetens the woody, unique taste of fresh asparagus. Adding oregano as an herbal element makes for a complex, yet overall light taste profile.

These flavors work nicely with light, white fish (such as halibut or sea bass), chicken, and pork preparations. Or take the approach that the application recipe does and add to this combination with protein-rich ingredients—making it a light meal all unto its own.

ASPARAGUS

Green is the most commonly available asparagus, but this vegetable also grows in purple and—the mildest flavored—white varieties. White asparagus stays this pale hue because it is denied light while grown. Asparagus is crispy when fresh, tender and delicious when cooked, and purées nicely. Blanching and shocking the asparagus, as in the application recipe, preserves the vegetable's color and texture.

ORANGE

The bright citrus flavor is a good contrast for the woody, herbal flavors of most vegetables. Here, this dish incorporates orange segments, juice, and zest to increase the orange flavor.

OREGANO

Don't confuse this herb with the milder and often mislabeled marjoram. Oregano is usually associated with Italian and French cuisine. It is easy to grow at home; try both the tangy Greek and aromatic Italian varieties. If using dried oregano, reduce usage to one teaspoon for every tablespoon called for in a recipe.

"WOODS IN SUMMER" RECIPE

Orange-Drenched Asparagus Salad

Wait for asparagus to come into season during the spring months. When blanching asparagus, always plunge it into ice water to stop the cooking process and preserve the vibrant green color. Adjust the cooking time by a minute in either direction depending on the size of your asparagus.

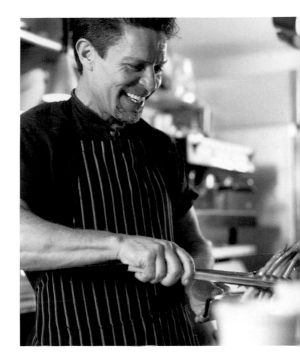

¼ cup (60 ml) olive oil

Juice and zest from 1 lemon

Juice and zest from 1 **orange**

1 small shallot, finely diced

2 tablespoons (30 ml) Champagne vinegar

2 tablespoons (8 g) **fresh oregano leaves**

Salt and pepper

1 pound (455 g) **fresh asparagus**, woody ends removed

1 **orange**, supremed (page 121)

2 eggs, hard boiled for 10 minutes, peeled, and chopped

¼ cup (36 g) almonds (preferably Marcona), toasted and salted

Whisk together in a small bowl the olive oil, lemon juice and zest, orange juice and zest, shallot, vinegar, oregano, salt, and pepper to taste. Set aside.

Prepare an ice bath by filling a large bowl with ice water. Bring a large saucepan of salted water to a boil. Add asparagus and cook for 2 to 3 minutes. Immediately plunge asparagus into the ice bath. After 5 minutes, remove asparagus from ice bath, shake off any excess water, and place on serving tray.

In the bowl with the vinaigrette, add orange segments, eggs, and almonds. Toss to combine, pour atop asparagus, and serve.

Yield: 4 servings

If you are lucky enough to live near a farm that allows you to pick your own asparagus, it's worth the trip. Snapping off the shoots from the ground is the next best thing to growing your own.

"ROOT VEGETABLES GONE SULTRY"

CARROT + FENNEL + WINE

Carrot's natural sweetness makes it versatile in cooking. This combination relies on its juice and the purity of its flavor. Its sweet lightness interacts well with the subtle fennel frond, giving it an anise undertone. Wine both incorporates and accentuates the flavors. This combination complements fish and chicken dishes nicely. Another option: Braise sliced carrot and fennel in white wine and serve it as a side dish.

CARROT

Consider the recipe when deciding which type of carrots to buy. Bunch carrots (those with greens still attached) work best for roasting and leaving whole, while bag carrots are great for juicing and chopping. Baby bagged carrots (not really "young" at all, but small from being trimmed and peeled) are a time-saver for puréed applications.

FENNEL

Fennel's most intense flavor comes when it is raw and crunchy. Sauté the root and simmer it to soften or roast it until it practically melts. Candy it for sweetness. Fennel seeds often flavor dishes, with the frond used as an herb.

WINE

All types of cooking call for wine of all shades and dryness. Generally, it cooks with other flavors to soften its impact or is used sparingly without cooking to heighten a dish's flavor profile. This application needs a dry white wine such as Chardonnay or Sauvignon Blanc that won't compete with the delicate flavors of carrot and fennel.

"ROOT VEGETABLES GONE SULTRY" RECIPE

Carrot Fennel Sauce

This sauce relies on carrot juice as its base. If you have a juicer with a vegetable attachment, by all means use it. If not, there are many high-quality bottled carrot juices from which to choose. Just be sure to buy one that contains no sugar or artificial ingredients—it will make a difference in the taste of your finished sauce. Serve the sauce over fish or chicken.

2 cups (475 ml) **carrot juice**

½ cup (120 ml) **dry white wine**

1 shallot, sliced thinly

½ cup (32 g) **fresh fennel frond**, large stems removed

¼ cup (60 ml) mirin (cooking sake)

1 teaspoon Sriracha sauce

Salt

1 tablespoon (14 g) unsalted butter

Add juice, wine, shallot, fennel, mirin, Sriracha, a pinch of salt, and butter to a large saucepan and cook over medium-high heat. Continue to cook until liquid reduces by half. Pass through a fine mesh strainer. Whisk in butter and serve immediately.

Yield: 1 cup (235 ml)

Because this sauce calls for the frond of fennel only, and because fennel is typically sold with the bulb, the bulb will remain. What a great opportunity to make Perfect-for-Fall Salad (page 117) to serve with your completed dish.

"SUNSHINE ON A COLD DAY"

GRAPEFRUIT + ONION + PARSLEY

Combining sweet and tart flavors, such as those in grapefruit, with the savory tang of onion and herbal freshness of parsley makes for complex and delicious effects. Cook this combination gently for a warm sauce, as in the application recipe. Or combine it raw, using red onion, for a chunky salad. The greatest variety is with the onion. Should you use a mild, strong, or sweet variety? Do you use it raw, sauté it, or caramelize it? These flavors are dynamite with seafood in particular but also with vegetables such as mushrooms and beets.

GRAPEFRUIT

Use both the grapefruit's segments and its juice for salads, reductions, and vinaigrettes. The pink and red varieties have a flavor similar to the white, but they add a nice color contrast to the other components. As a substitute for grapefruit, try the pomelo (sometimes called Chinese grapefruit), which is larger in size but more mellow in flavor.

ONION

Part of this combination's appeal lies in contrasting sweet citrus with tangy onion, so for maximum effect, stay away from the sweet varieties or red onions (unless you are using them raw). The application recipe calls for pearl onions, the perfect small, mild morsels to serve with grapefruit segments. They do take some time to peel—make this easier by cutting off the ends, quickly blanching them in boiling water, and popping them out of their skins.

PARSLEY

Two parsley varieties dominate the market: curly or Italian flat leaf. Choose the flat-leaf variety. It is more fragrant and less bitter than the curly variety. It also looks better in food, and has a more upscale appearance (remember, food that looks better actually tastes better too).

"SUNSHINE ON A COLD DAY" RECIPE

Savory Grapefruit Sauce

Pair this sauce with seared scallops or sea bass for a flavorful dish.

6 to 8 **white pearl onions** (other colors will taste good,
  but white aesthetically looks better)

2 tablespoons (8 g) **Italian flat-leaf parsley**, leaves only, divided

1 teaspoon (1 g) fresh thyme leaves

1 tablespoon (15 ml) canola oil

1 shallot, finely diced

¼ cup (60 ml) dry white wine (such as a New Zealand Sauvignon Blanc)

2 **grapefruits**, supremed (page 121)

½ cup (120 ml) **grapefruit juice** (fresh squeezed)

¼ cup (85 g) honey (preferably lavender)

Salt and pepper

1 tablespoon (15 ml) almond oil

2 tablespoons (28 g) unsalted butter

Cut off both ends of the pearl onions and blanch for 1 minute in a pot of salted boiling water. Drain and squeeze the inner onion out of its skin.

Chop together 1 tablespoon (4 g) of parsley leaves with the thyme. Set aside the remaining 1 tablespoon (4 g) of whole parsley leaves.

In a large skillet, heat canola oil and sauté shallot over medium heat. Cook until softened and translucent. Squeeze juice from the two fresh grapefruits. Deglaze pan with the wine, then add grapefruit juice (not segments) and honey. Adjust seasoning with salt and pepper and bring to a simmer. Whisk in almond oil and butter add chopped parsley and thyme.

Just prior to serving, add onions, grapefruit segments, and whole parsley leaves.

Yield: 2 cups (475 ml)

Grapefruit is in season during winter—perfect for a bright, colorful dish during those drab, cold months.

"TROPICAL FRUITS GONE NUTS"

AVOCADO + MANGO + NUT

Avocado and mango both have a rich, luscious texture but with different flavors—nutty and mild for the former, sweet and tropical for the latter. Together, they are a dynamic, rich-sweet pairing made even better by adding crunchy nuts. Because the fruits generally don't get cooked, these flavors are well suited to salads. Also pair them with seafood or chicken.

The application recipe makes a flavorful topping for crab cakes. An alternative includes a spinach salad with cubed avocado, mango, and roasted macadamia nuts

AVOCADO
Avocados tend to oxidize quickly, so make sure that if you prepare them ahead of time, you add lime or lemon juice, which slows down the darkening process. For easy eating on the go—and much healthier than anything you would get at a drive-through—use a spoon to scoop bites directly from the avocado shell.

MANGO
Its juice is a thick mouthful. Use it fresh for clarity of flavor in salads and relishes. Cooking softens the texture and brings out its natural sugars. When dried, it becomes sweeter and its shelf life extends. If you end up with a not ripe mango, put it in a paper bag with a banana and give it a day or two at room temperature. Or consider roasting it with orange juice to bring out the sugars.

NUT
Toasting and roasting nuts intensifies their flavors. Eat them raw out of your hand, turn them into butters or brittle, or use them to add a textural note to a complex dish. For this pairing, think of pine nuts for a bold entrance or macadamia for a less dominant flavor. Always buy your nuts raw so you can control the roasting and seasoning.

"TROPICAL FRUITS GONE NUTS" RECIPE

Tropically-Topped Crab Cakes

These crab cakes are packed with crab, not with bread, as they often are served in restaurants. Be sure to buy real crabmeat (not imitation). Save yourself effort by buying the meat only, not the shells.

Avocado-Mango Topping

1 **mango**

1 clove garlic, finely chopped

1 shallot, finely chopped

6 sprigs fresh cilantro

6 sprigs fresh mint

Juice of 1 lime

1 teaspoon (5 ml) rice vinegar

1 teaspoon (5 ml) simple syrup (page 183)

1 **avocado**, diced

1 tomato, diced

¼ cup (35 g) **pine nuts**, toasted

Sriracha sauce (optional, to taste)

Crab Cakes

1 shallot, finely chopped

2 cloves garlic, finely chopped

2 tablespoons (13 g) leek (the white part only), finely chopped

1 tablespoon (15 ml) olive oil

1 pound (455 g) crabmeat, chopped

Juice of 1 lemon

1 tablespoon (11 g) Dijon mustard

½ cup (60 g) bread crumbs (panko preferably), divided

1 egg, lightly beaten

¼ cup (60 g) aioli or mayonnaise

Salt and pepper

2 tablespoons (28 g) unsalted butter

To prepare avocado-mango topping: Dice the mango into a 1-inch (2.5 cm) dice, reserving the scraps or slivers. Place these scraps (should be around ¼ cup [40 g]) into a blender. Add the garlic, shallot, cilantro, mint, lime juice, vinegar, and simple syrup. Blend until smooth. This is your binder for the topping.

In a large bowl, combine diced mango, avocado, tomato, pine nuts, and Sriracha (if using). Add the mango purée and stir to incorporate. Cover and refrigerate until using, up to 2 hours before serving.

To prepare crab cakes: Sauté shallot, garlic, and leek in olive oil until tender. In a large bowl, add sautéed vegetables to crabmeat, lemon juice, mustard, ¼ cup (30 g) bread crumbs, beaten egg, and aioli. Mix together and season with salt and pepper. Using your hands, form mixture into small cakes. Coat with remaining bread crumbs.

Heat butter over medium heat in a skillet. Cook crab cakes until golden brown and warm on the inside. Don't be tempted to raise the heat or you may have a nice golden sear on the outside but cold crab on the inside. Top with avocado-mango salad and serve immediately.

Yield: 4 servings

If preparing crab cakes for guests, you can sear them up to 1 hour ahead of time and leave them at room temperature. Place in an oven heated to 350°F (180°C, or gas mark 4) for 5 to 8 minutes to warm before serving.

"A BURST OF BRILLIANCE"

PARSLEY + GARLIC + LEMON

Think of the possibilities for this combination of herbal freshness, tangy bite, and tart lemon. When combined, they create a full, bright, exciting, light flavor.

Bring the flavors together as the application recipe does with an incredibly versatile gremolata (an Italian garnish). A gremolata uses the lemon zest only; for an alternative, make vinaigrette by mixing the juice from the lemon and oil. Use the combination to season baked meats and fish as well, with preserved lemons, garlic, white wine, butter, and a garnish of fresh parsley. These flavors are well suited to braised and grilled meats (especially veal and lamb), seafood, potatoes, and vegetables such as cauliflower, broccoli, and summer squash.

PARSLEY

Two parsley varieties dominate the market: curly or Italian flat leaf. Choose the flat-leaf variety. It is more fragrant and less bitter than the curly variety. There is no room for dried parsley in this combination, and because grocery stores carry fresh year-round, you have no excuse. If you have a sunny window in your kitchen, consider growing your own parsley in a pot so you get the freshest, most flavorful product.

GARLIC

Avoid garlic-flavored salts or powders, which have their place in spice blends, but are a far cry from the real thing in cooking. A head of fresh garlic will keep in a dark, cool spot for a couple of weeks, and once you have it on hand, you are more likely to use it. Sauté it for an alluring aroma and softened taste, leave it raw for potent heat, or roast it for a mellow sweetness.

LEMON

In this combination, use the lemon's finely grated zest. To do the job easily and without including the bitter white pith, you really need a micrograter. Be sure to wash the lemons well—even more so if they have a wax coating. For an interesting variation, try substituting lime or orange zest.

"A BURST OF BRILLIANCE" RECIPE

Gregarious Gremolata

Okay, technically gremolata does not talk, but its flavors are definitely outgoing. It can be the life of your party with its bright burst in each bite. Here, it adds a refreshing note to roasted broccoli.

1 head broccoli, cut into florets
2 tablespoons (30 ml) olive oil
Salt and pepper
2 tablespoons (8 g) chopped **flat-leaf parsley**
1 clove **garlic**, minced
Zest and juice from 1 lemon, divided

Preheat oven to 425°F (220°C, or gas mark 7). In a bowl, toss together broccoli, olive oil, salt, and pepper to taste. Spread out on a sheet pan and roast for 20 minutes until tender and browned.

While the broccoli is in the oven, prepare the gremolata by combining the chopped parsley, garlic, and lemon zest (not the juice). Season with salt and pepper.

Once the broccoli is cooked, sprinkle it with the lemon juice, then the gremolata. Serve warm or at room temperature.

Yield: 4 servings

Feel free to substitute your favorite vegetables for the broccoli—green beans, cauliflower, and brussels sprouts all make nice choices. Also, add some olive oil to the dry gremolata blend to coat vegetables or to use as a rub for meats.

"CAPTIVATINGLY COOL"

PEAR + CUCUMBER + LEEK

This combination's nice balance of texture and flavor excites the palate. The sweet flavor and buttery texture of pear meet the refreshing, crisp cucumber and intermingle with the soft oniony heat of leeks.

The application recipe takes advantage of variations for this combination, resulting in a crunchy raw salad and a sautéed, puréed broth. Feel free to use them separately or, as the recipe indicates, together with fish. In addition to fish preparations, these flavors work well with oysters.

PEAR

Use a ripe, versatile pear such as Anjou or Comice for cooking and puréeing. The recipe calls for Asian pear, quite different from most pears with its apple shape and flavor and crisp, firm flesh (it is supposed to be that way—it's not underripe). If you cannot find Asian pears, use a more common pear variety. Just realize that the texture of the salad will change; you may consider using an underripe pear for additional texture.

CUCUMBER

When making a purée, as in this recipe, don't waste your time peeling and seeding the cucumber. If it has a wax coating, wash thoroughly before using but purée the whole vegetable, seeds and all. Simply strain the mixture for a full-flavored product. Cucumber purée also makes a nice base for beverages—a little sparking water or a shot of gin both offer tasty options.

LEEK

Although the light, white portion of the leek is most commonly used, once you remove the tough exterior leaves, you can certainly use the darker portion. Save it to add to stocks or cut into a small dice for sautéing. Know that the inner leaves are a magnet for dirt and sand, so slice lengthwise and always wash really well.

"CAPTIVATINGLY COOL" RECIPE

Braised Halibut with Warm Cucumber Broth

This recipe prepares both a delicate cucumber broth and a crunchy salad to serve with fish.

Cucumber Broth

½ cup (15 g) spinach leaves

½ cup (48 g) mint

½ cup (30 g) fresh parsley leaves

2 tablespoons (30 ml) vegetable oil

1 **leek**, darker ends diced, light-green section reserved for salad

1 small onion, diced

2 stalks celery, diced

½ large **cucumber**

1 cup (235 ml) water

¼ cup (60 ml) semidry white wine

Juice and zest from 1 lemon

1 teaspoon (5 ml) rice vinegar

Salt and pepper

Fish

4 pieces white fish, such as halibut or sea bass, 6 to 8 ounces (170 to 225 g)

Salt and pepper

2 tablespoons (28 g) unsalted butter

2 tablespoons (30 ml) semidry white wine

Salad

1 **leek**, light-green section

½ **cucumber**, peeled and seeded

1 **Asian pear**, peeled and cored

12 mint leaves

12 celery leaves

To prepare cucumber broth: In a saucepan of boiling, salted water, blanch spinach, mint, and parsley for 30 seconds. Immediately plunge into an ice bath; remove to drain on paper towels.

In a skillet over medium heat, heat oil and sauté leek, onion, and celery. Cook until tender and translucent. Place in a blender and add cucumber, water, wine, and lemon juice and zest. Add blanched spinach, mint, and parsley and blend for 2 minutes or until fully incorporated. Strain through a fine mesh strainer, add vinegar, and season with salt and pepper.

To prepare fish: Preheat oven to 350°F (180°C, or gas mark 4). Place fish on a sheet pan and season with salt and pepper. Add a large dab of butter to each piece of fish and then drizzle with wine. Cook for 12 minutes, then squeeze fish to determine whether it is cooked through. Cook for an additional 2 to 3 minutes, if necessary.

To prepare salad: While the fish cooks, slice leek, cucumber, and Asian pear into matchsticks. Combine with mint and celery leaves in a small bowl.

To serve: Warm cucumber broth and spoon into the bottom of a shallow bowl. Top with braised fish and salad. Serve immediately.

Yield: 4 servings

🍴 **This combination, perfect for a sweltering summer night, has light, cool flavors that are incredibly satisfying.**

"AN UNFAMILIAR BREEZE"

GRAPE + SAFFRON + NUT

This is a combination of exotic complexity. The honey-like aroma and bright color of saffron make the stimulating, crisp grape even more alluring. Nuts provide an additional layer of flavor and a crunchy edge that unifies the combination.

The application recipe uses the ingredients with fish for a crunchy pistachio coating and grape-saffron salsa. Another option: Create a green salad with chicken, grapes, and saffron vinaigrette. These flavors are well suited to fish, chicken, dates, and couscous.

GRAPE

Sweet with a bit of acidity, any grape variety (green, red, or purple) will do nicely in this combination. Consider the look of your overall dish and choose a color that contrasts with the other components. Look for less common varieties such as Muscat or Concord in late summer to early fall at farmers' markets. Remove seeds from grapes before using—a relatively simple but time-consuming task.

SAFFRON

Yes, saffron is expensive, but when you consider that it takes 75,000 blossoms of saffron crocus flower to produce one pound (455 g) of saffron threads, you can understand why! Threads of saffron add exotic flavor to any dish, and a little goes a long way. In fact, be careful not to add too much to your cooking or you may end up with a medicinal-tasting dish. Always buy the threads, not powder, and store in the refrigerator for maximum shelf life.

NUT

Because the saffron spice has an exotic flavor, choose a nut with its own uniqueness—pistachios, cashews, and macadamias work nicely. Almonds, though not particularly distinct, have an accommodating mild flavor. Stay away from walnuts or hazelnuts, which may interfere with the saffron flavor profile.

"AN UNFAMILIAR BREEZE" RECIPE

Pistachio-Crusted Halibut with Grape-Saffron Salsa

This is a great crispy fish dish that is full of flavor but not heavy. Feel free to substitute your favorite fish fillet.

Salsa

1 mild red onion, finely chopped

2 cloves garlic, finely chopped

1 mild fresh red chile pepper (such as
  paprika), seeded and finely chopped

2 tablespoons (8 g) chopped fresh Italian
  parsley leaves

24 **seedless green grapes**, halved

2 Roma tomatoes, diced

½ cup (65 g) diced jicama

1 teaspoon (1 g) **saffron threads**,
  lightly toasted

3 tablespoons (45 ml) sherry vinegar

¼ cup (60 ml) olive oil

Fish

Salt and pepper

4 halibut fillets, 6 to 8 ounces
  (170 to 225 g)

⅓ cup (47 g) cornmeal

¼ cup (31 g) **toasted pistachios**, ground
  in food processor

3 tablespoons (42 g) butter, for frying

2 eggs

Combine all of the salsa ingredients and allow to infuse while preparing fish.

Lightly salt and pepper both sides of the fish. Combine cornmeal and pistachios in a shallow bowl and season with salt and pepper. Beat eggs in another large shallow bowl.

Heat butter in a large skillet over medium heat. Dip each piece of fish into the beaten eggs and then coat in the crumb mixture. Fry on each side for 3 to 4 minutes until golden brown. Serve immediately topped with salsa.

Yield: 4 servings

🔆 This combination and recipe could work equally well with chicken breasts. Pound them out between two pieces of parchment paper to make thinner and to take advantage of the crispy coating.

"THE BEE'S KNEES"

APRICOT + HONEY + THYME

The small, sweet, aromatic apricot stars in this simple gathering of ingredients. Both the honey and the cooking process accentuate the apricot's natural sugars. Adding the herbal thyme keeps this from dropping off the "overly sweet" edge. While a natural in desserts, this combination also works for savory applications, especially with pork, game meats, and cheeses. Perhaps used as a honey-thyme marinade for grilled apricots and pork tenderloin? What about as an infused jam?

APRICOT

This stone fruit has a relatively firm flesh that holds up nicely to heat. When cooked (by sautéing, roasting, braising, or grilling), its slightly musky taste sweetens and its texture softens. In this combination, to better incorporate the honey and thyme flavors, you must cook the apricot in some way.

HONEY

Artisanal honeys abound, distinguished by what the bees are pollinating. If you can find it, thyme honey would be the obvious choice for this combination. If not, use a mild honey such as clover or tupelo. Or infuse your own by plunging your choice of herb into a small jar of honey and allowing it to sit for at least five days.

THYME

Because bees are fond of thyme, this is a natural pairing with honey. Thyme is an easily grown herb that has a mild flavor and blends well with other herbs and ingredients. Garden thyme is the most readily available variety, but look for lemon thyme as well, prized for its lemon aroma (which pairs especially nicely with fish).

"THE BEE'S KNEES" RECIPE

Warm Apricots with Thyme-Honey Reduction

This simple dessert preparation can easily be embellished with a dollop of crème fraîche or vanilla ice cream.

½ cup (170 g) **honey** (preferably **thyme**)

Juice of 1 lemon

1 large sprig **fresh thyme**

6 **apricots**, halved and pitted

1 tablespoon (2.4 g) **fresh thyme**,
    leaves only

In a large skillet, combine the honey, lemon juice, and thyme sprig. Bring to a simmer over medium-high heat and let it reduce slightly.

Place the apricots in the skillet cut-side down, and cook for 2 minutes or until tender. Turn apricots over and cook for an additional 1 minute. Remove apricots from the syrup and transfer to a plate.

Remove the thyme sprig from the syrup and reduce the syrup for 1 minute. Drizzle over apricots and sprinkle thyme leaves on top. Serve immediately.

Yield: 4 servings

This simple dessert goes down nicely with a glass of sweet dessert wine, which often has apricot overtones. Look for a Muscat or a late-harvest Riesling or ask your local retailer to recommend one that pairs well with apricots.

"RELEASE THAT RHUBARB!"

RHUBARB + GINGER + LEMON

Free rhubarb from under all of those pies. This vegetable may look like red celery but it is really quite tart. Letting it rub shoulders with spicy ginger creates a flavorful contrast made even brighter by the tart accent of lemon.

Here it is cooked to become a flavor-packed jam. Other variations include a more lemon-forward flavor such as a lemon mousse with rhubarb-ginger purée. Or consider baking it into a sweet bread using fresh rhubarb, ground ginger, and lemon zest.

RHUBARB

Rhubarb is in season in the spring through the summer, so look for the freshest, most flavorful product then. Rhubarb releases much of its moisture during cooking. When heated with sugar, it turns into syrup. When cooked for long enough, it breaks down into a pulp, so if you want it to retain its shape, keep a close eye on it.

GINGER

Ginger's zing can be cut from the fresh root, extracted in its juice, crystallized, candied, or dried and ground. The flavor of dried ginger is very different than fresh ginger, but if you have to substitute, use ½ teaspoon ground ginger for 1 tablespoon (8 g) grated fresh. Ground is generally better suited to baking applications. In this application recipe, candied ginger imparts the ginger flavor and sweetens the combination.

LEMON

With rhubarb as the lead flavor, lemon as an acid component makes the flavors more dynamic. Zest adds flavor, and the segments, if used, provide even more lemony goodness. Never use lemon juice from a bottle; it just isn't the same intensity.

"RELEASE THAT RHUBARB!" RECIPE

Tart and Tangy Compote

This recipe produces a thick, jam-like compote. Use it with savory foods, as in a relish for duck or scallops or as a dessert with ice cream, cheesecake, or even pancakes.

4 cups (490 g) diced **rhubarb**
3 cups (600 g) sugar
¼ cup (35 g) finely diced **candied ginger**
Juice and zest from 1 **lemon**

Combine all ingredients in a saucepan and let sit for 15 minutes. Bring to a simmer over medium heat and allow to cook, stirring often, until it thickens and the rhubarb breaks down.

Remove from the heat and skim off any foam from the top. Cool, cover, and refrigerate for up to 1 week.

Yield: 2 cups (640 g)

Think of this jam as a jumping-off point for other applications. Purée it for many alternatives: a smooth sauce for ice cream, a cold soup (with more juice), or a spritzer with sparkling water or Champagne.

CHAPTER 5

Sweet & Sour

Imagine the sugar-rush feel of sweetness when a sugar cube dissolves on your tongue or the puckering sour effect of sucking on a lemon wedge. Put together this childhood combination and you get the principles at work in this chapter. Most world cuisines use sweet and sour elements; when combined correctly, they create a bit of culinary magic both stimulating and satisfying. Achieving balance between the two is key to creating a dynamic dish.

Sweet-and-sour pairings aren't always done well. Take a cheap version of a Chinese sweet-and-sour chicken, for example. Often the sweet element overwhelms the viscous, gummy sauce, from reliance on a low-quality and too-light use of sour. Using high-quality ingredients (not necessarily gourmet, mind you) is always important in creating tasty flavor combinations, particularly with sweet-and-sour applications.

Choose a naturally sweet ingredient (such as raisins, for example) or use additional sweeteners such as granulated sugar, brown sugar, or honey for the sweet factor. For a sour taste, use one of the many varieties of vinegar (rice, Champagne, malt, red wine, balsamic, etc), an element of citrus, or a naturally sour ingredient (such as persimmon). The two should interact such that one does not overpower the other and destroy the delicate balance.

A section of this chapter focuses on pickles. Bringing salt and acid to an ingredient stabilizes the aging process, giving the food a longer shelf life. Besting that advantage, the pickles' flavor and crunch steal the show. Use these condiments for everything from snacking to salads and sandwich toppings. At our gourmet burger pub, we use the Turmeric-Pickled Cauliflower (page 157) with the falafel burger. Choose a rich or sweet pairing with a pickle to produce a complex flavor outcome.

If you remember to keep the sweet and sour in balance (as these recipes demonstrate) and use good ingredients, you will end up with wonderfully interesting sweet-and-sour dishes—kind of like when, as a child, you combined the sugar cube with the lemon wedge.

"MIDDLE EASTERN CRUNCH"

CAULIFLOWER + TURMERIC + VINEGAR

Cauliflower has a unique but mild taste that blends well with the lovely-hued, tangy turmeric. The vinegar component balances the combination or as in the example that follows, allows for a dynamite pickle when increased in proportion to the other ingredients. These flavors work nicely with light meat such as chicken or turkey, tofu, and other vegetables such as carrots and parsnips.

CAULIFLOWER

This vegetable gives a great crunch when eaten raw for crudités; stir-frying it retains its crispness and reduces the flavor. Baking and boiling softens its texture and allows for the insertion of other flavors. Once cooked, it can be puréed into a smooth side dish.

TURMERIC

This bright yellow spice most commonly acts as a base for curries, but it also seasons meat and vegetables in Middle Eastern cuisine. Use it as an accent, but resist the urge to let it dominate—unless you are deliberately trying to dye your teeth yellow!

VINEGAR

For this combination, stick with light vinegars—white or rice both work well, and their colors won't interfere with turmeric's vibrant hue. To let it play the starring role, increase its proportions and make a pickle.

"MIDDLE EASTERN CRUNCH" RECIPE

Turmeric-Pickled Cauliflower

Pickles are a great way use up a garden bounty and eat more vegetables. Beyond snacking, they taste great in salads, with dips or hummus, or atop a sandwich (such as the Edamame Falafel, page 209).

Juice of 1 lemon

Juice of 1 lime

Zest from 1 orange

1 fresh jalapeño pepper, sliced
   lengthwise

2 cloves garlic, smashed with the back
   of a knife

1 inch (2.5 cm) fresh ginger, peeled
   and rough cut

2 tablespoons (36 g) salt

1 tablespoon (6 g) pickling spice

1 dried bay leaf

1 cinnamon stick

¾ cup (175 ml) **white vinegar**

2 cups (475 ml) water

½ cup (120 ml) white wine

¼ cup (50 g) sugar

2 tablespoons (40 g) honey

Pinch saffron

1 tablespoon (7 g) **turmeric**

1 large head **cauliflower**, cut into small florets

Mix all ingredients except for the cauliflower in a large stockpot and bring to a boil. Add cauliflower and bring back to a boil. Remove from heat and allow to come to room temperature. Keep covered in its liquid in the refrigerator.

Yield: 2 cups (200 g)

Eating cauliflower offers many health benefits: maintaining a healthy cholesterol level, increasing fiber intake, and providing a rich source of folate, to name a few. A pickle like this one gives you more opportunities to enjoy this delicious, healthy ingredient.

"PROVOCATIVE PEACHES"

PEACH + GINGER + VINEGAR

A Persian dish called torshi holu inspired this combination and recipe. Persians place pickles in high esteem for their positive effect on body functions. You can make them simply because they are delicious.

The sweet peach in this pickle recipe adds a pleasant fruit-driven flavor to an otherwise spicy-sour blending. The tang of fresh ginger is a natural partner to peach, giving the combination a complex edge. You need the big boost of vinegar to create the pickle, but scaling the liquid back could result in a dynamite ginger vinaigrette to serve over sautéed peaches and fresh spinach.

PEACH

Determine ripeness—critical when eating peaches raw—by pressing lightly with your thumb. The fruit should give slightly under this pressure. Peaches are cooked to increase their natural sugars, resulting in a softened consistency. In this application, feel free to experiment with white peaches for a slightly different effect. And if the peaches aren't ripe, substitute other ripe stone fruit such as nectarines, apricots, or plums.

GINGER

This tuber, with its potent juice and flavor-filled fresh root, is both distinct and versatile. Cooking ginger mellows its impact and allows its incorporation with other flavor profiles. Pickled ginger retains a crunchy texture, while candied or crystallized ginger has a chewy, sweet taste. Use fresh for this combination, for purity of flavor, but substituting 1 tablespoon (6 g) of powdered ginger works as well.

VINEGAR

With the flavors of peach and ginger, stick with a light vinegar. Experiment with Champagne or rice vinegar for different effects on the finished dish.

"PROVOCATIVE PEACHES" RECIPE

Perfectly Pickled Peaches

These peaches traditionally are served with kebabs and fried meat in the Persian culinary scene. They also add zing to rice dishes, couscous, or orzo preparations.

2 inches (5 cm) **fresh ginger**, peeled and chopped

2 cups (475 ml) **white wine vinegar**

2 tablespoons (10 g) coriander seeds, toasted slightly

1 whole bulb garlic, separated and peeled

1 pound (455 g) **fresh peaches**

½ teaspoon ground red pepper (cayenne preferred)

¼ cup (50 g) sugar

1 teaspoon (2 g) freshly ground pepper

Combine all ingredients in a large saucepan or stockpot. Bring to a boil and cook for 5 minutes. Allow to come to room temperature in the liquid, adding additional vinegar if the liquid does not entirely cover the peaches. Cover and keep refrigerated until serving or for several weeks.

Yield: 6 to 8 peaches

Summer is peach season, so take advantage of the bounty to try this recipe. For longer storage, seal in sterilized glass jars. Then you can enjoy these flavorful peaches during the cold winter months.

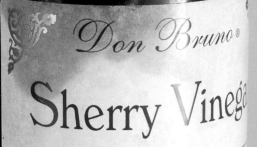

"NOT SO COOL, CUCUMBER!"

CUCUMBER + JALAPEÑO + VINEGAR

Cucumber pickles are so ubiquitous in American cuisine that you may not even notice one on the plate with your hamburger. In this combination, the cucumber pickle gets attention! There's the initial crunch of the pickle, quickly finished with the lingering spiciness from the fresh jalapeño. The boost of vinegar is needed to actually make this a pickle. By varying the proportions of these components, you could easily create a spicy vinaigrette for a fresh cucumber salad.

CUCUMBER

Eat them raw, slice them up and serve with dip, or chop them and throw into salads. For this combination, the cucumber needs to keep its crisp bite to effectively deliver the spicy component of the jalapeño, so we recommend keeping it raw or cooking it gently.

JALAPEÑO

These small green peppers pack a spicy wallop. Use them sparingly raw or jarred to add heat. Roasting jalapeños softens their flavor and texture considerably and allows use in a larger proportion to the cooling cucumber. To sneak a little spicy flavor into a dish, include in your recipe a bit of the liquid from a jar of sliced jalapeños.

VINEGAR

Vinegar is a component essential to the pickling process. White vinegar works best for this combination. To tweak it for a vinaigrette, decrease the vinegar and add oil to finely minced cucumber and jalapeños.

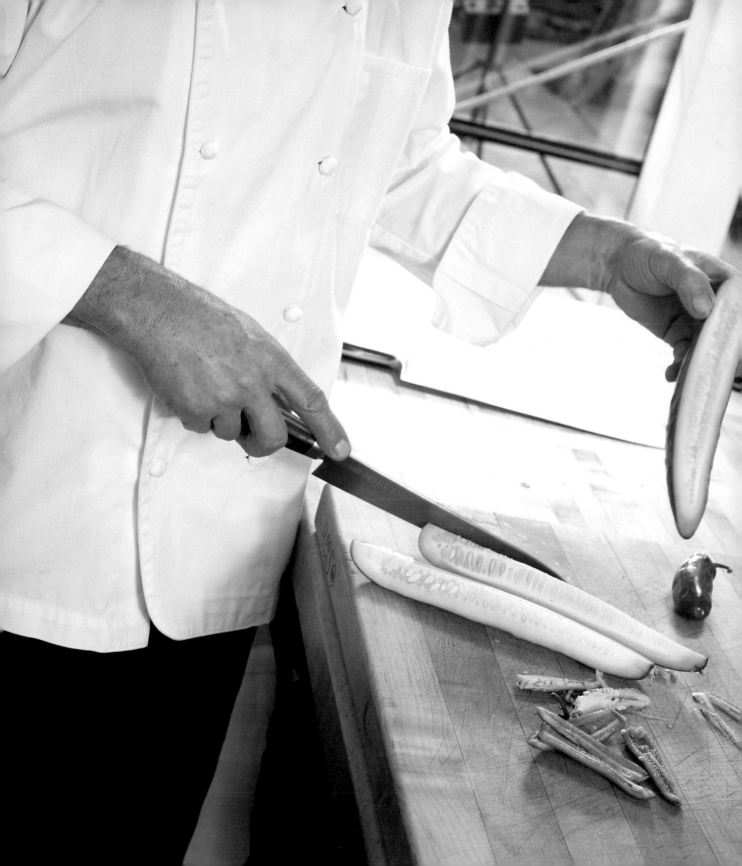

"NOT SO COOL, CUCUMBER!" RECIPE

Pickles with a Punch

Surprise your kids when you put these spicy spears next to their ham sandwiches.

1 to 2 whole **jalapeño peppers** (depending on heat preference)
2 medium **cucumbers**
1 teaspoon (2 g) freshly ground pepper
2 tablespoons (36 g) salt
¼ cup (50 g) sugar
¼ cup (16 g) rough-chopped fresh dill
1 clove garlic
¼ cup (4 g) rough-chopped cilantro
2 tablespoons (30 ml) white wine
1½ cups (355 ml) **white vinegar**
2 tablespoons (12 g) pickling spice
2 cups (475 ml) water
Pinch red pepper flakes
2 teaspoons (10 ml) Sriracha sauce

Slice peppers in half lengthwise and remove inner core and seeds. Cut in half again, so each pepper is in quarters. Set aside.

Slice cucumbers in half lengthwise. Slice each half into thirds lengthwise. Cut each long piece into desired length of spears. Set aside and prepare pickling liquid.

Place pepper quarters and all remaining ingredients (not including the cucumbers) in a large stockpot and bring to a boil. Once boiling, add cucumber spears and remove from heat. Allow to come to room temperature. Keep covered in its liquid in the refrigerator for several weeks.

Yield: 24 to 30 pickles

This is a great recipe to prepare mid-summer when fresh cucumbers and peppers are abundant. Support your local farmers' market or, better yet, grow your own.

"SWEET INTRIGUE"

APRICOT + BRANDY + CARDAMOM

The delicate, honeyed sweetness of fresh apricot gets a burst of power with the smooth-ness of brandy and the savory-sweet notes of cardamom. The interplay of these flavors produces a delightful mouthful that works with different cooking techniques.

Try the application recipe for a quick, flavorful apricot pickle, sauté and simmer the apricot and brandy for a sauce for pork chops, or tuck brandied apricots into sweet car-damom bread for a baking application. Make this combination in savory or sweet dishes and with foods including chicken, pork, ginger, and ice cream.

APRICOT

This stone fruit has a relatively firm flesh that holds up nicely to heat. When cooked (by sautéing, roasting, braising, or grilling), its slightly musky taste sweetens and its texture softens. Dried apricots can be cooked in brandy to plump and added to sauces or cardamom-sweetened syrups.

BRANDY

Fruit-based brandies work nicely with these flavors—try an eau de vie such as Pear Williams or Calvados for an apple brandy. These can be a bit pricey, but they are high quality for sipping on many occasions. Don't use anything described as fruit-flavored brandy, which typically contain high quantities of artificial ingredients. For great California brandy made in the Cognac tradition, try those by Germain-Robin.

CARDAMOM

When using whole pods in beverages (usually to infuse a liquid), be sure to break them open to release the inner seeds and flavor. Using ground cardamom imparts a significant amount of flavor, so start with a small amount and add more to taste. Try to use pods when possible. When ground cardamom is called for, it's best to toast your own pods and grind in a spice grinder.

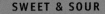

"SWEET INTRIGUE" RECIPE

Apricot Pickle for All Occasions

During the short summer months when apricots are in season, stock up and make this pickle to enjoy for up to six weeks. It tastes delicious with pork or chicken, on a cheese tray, or as a snack with a glass of wine.

1 cup (235 ml) cider vinegar

1 cup (235 ml) orange juice (preferably fresh squeezed)

½ cup (120 ml) mango juice

¼ cup (50 g) sugar

¼ cup (60 ml) maple syrup

8 to 10 **cardamom pods**, crushed open with the back of a knife

Zest of 1 orange

1 inch (2.5 cm) fresh ginger, peeled and minced

8 **apricots**, quartered and pitted

½ cup (120 ml) **Calvados apple brandy** (or substitute a nice brandy)

In a large saucepan, bring the vinegar, juices, sugar, and maple syrup to a boil, stirring to dissolve the sugar. Add the cardamom and simmer over low heat for 5 minutes. Remove from heat and add the zest and ginger. Allow to cool to room temperature and then add the apricots and brandy. Cover and refrigerate for up to 6 weeks.

Yield: 4 cups (655 g)

ii Before eating these apricots, give them a few days for the flavors to infuse. Your patience will be rewarded by deeper, more satisfying flavors. Substitute plums or pluots (plum-apricot hybrids) if you have them.

"BLUSHING SCARLET"

RADISH + PEPPERCORN + ROSE HIP

This combination comes alive with flavor! Radish has a sharp, spicy crunch accented by the peppercorn's heat and given a fruity-tart edge with the rose hip.

The application recipe prepares a fabulous, vibrantly colored pickled radish. For an alternative, slice radish thinly on a mandoline and top with a light white pepper–rose hip vinaigrette. Try this combination with seafood such as scallops and shrimp, tofu, and spicy greens such as arugula.

RADISH

Beyond the small, red round, or oblong radishes, look for other varieties. Easter radishes—not a variety as much as an assortment of colors—come in hues including purple, white, and pink. Daikons are large radishes commonly used in Asian cooking that offer a milder flavor. Black radishes, with black skin and white flesh, are distinctly spicier than other radishes. For maximum flavor and crunch, use raw. Radishes also braise or roast nicely for a softened texture and flavor.

PEPPERCORN

In this combination, try pink peppercorns for their fragrance, delicate flavor, and aesthetically compatible color. Green peppercorns also provide a color contrast and mild flavor (achieved by picking when they are underripe). Whichever peppercorn you choose, use whole in liquid applications or grind it yourself for maximum flavor.

ROSE HIP

Rose hip is the fruit of the rose plant, which we don't often see because the rose flower is usually harvested and its plant pruned back. Rose hips offer a spicy-tart flavor comparable to cranberry. Use them fresh or dried, the latter of which is the most readily available, usually in health food stores. If you have fresh, use twice as much as dried.

"BLUSHING SCARLET" RECIPE

Revolutionary Radish Pickle

Not sure what to do with that bunch of radishes from your CSA (Community Supported Agriculture) bag of produce? Pickle it! This recipe will win over any reluctant radish eaters. Use on vegetable crudités, in salads, and in Asian rice dishes.

1 bunch **radishes**, stemmed and
    quartered
½ cup (120 ml) rice wine vinegar
1 teaspoon (6 g) salt
2 tablespoons (25 g) sugar
1 teaspoon (2 g) whole **peppercorns**
(preferably pink)
1 teaspoon (1 g) **dried rose hips**

Place radishes in a bowl. In a small saucepan over medium heat, stir together the vinegar, salt, sugar, peppercorns, and rose hips. When the sugar dissolves and just begins to simmer, remove from heat and let cool for 10 minutes. Pour over radishes and allow to come to room temperature. Cover and refrigerate for 8 hours. These pickles will keep in their liquid in the refrigerator for up to 2 weeks.

Yield: 1 cup (116 g)

Okay, we may as well tell you that these pickles produce a pungent aroma. So be sure to keep them covered in the refrigerator and don't be surprised when you open the container for the first time. Be brave—these are delicious.

"SWEET AND SOUR HARMONY"

RAISIN + MUSTARD SEED + VINEGAR

This sweet-and-sour application begins with the raisin, which gains intense sweetness when the grape's natural sugars come out during the drying process. The mustard seed is incorporated as a seasoning, imparting a mild aroma on the raisin. The vinegar gives the combination its sour element, resulting in a wonderful sweet-and-sour condiment. Use with roasted chicken, pork chops, lamb, pâté, or ethnic rice dishes. Or grind the mustard seed to make a homemade tangy-sweet spread with raisins and a pinch of cinnamon.

RAISIN

This recipe calls for golden raisins, which tend to plump better than the dark variety and are more versatile with various meats. Golden raisins sometimes come from the same grape variety as their darker brethren (sometimes not), but they are oven-dried (rather than sun-dried) and are treated with sulfur dioxide to prevent darkening. Use the soft and sweet sultana if you can find them. If you substitute dark raisins, consider using a stronger vinegar variety.

MUSTARD SEED

This application calls for toasting the seeds, a step that deepens their flavor and aroma. These seeds come in yellow, brown, and black, and here, color matters; color affects the flavor of the seed, so follow the recipe's ingredient suggestion. To make your own mustard, grind the seeds with honey and raisins, white wine vinegar, and white grape juice.

VINEGAR

Let your choice of raisin guide your choice of a vinegar. For light raisins, select a white wine or Champagne vinegar. For darker raisins, go with a sherry or red wine vinegar.

"SWEET AND SOUR HARMONY" RECIPE

Dynamite Raisin Compote

This is just the condiment to produce a "wow" factor—even on simple dishes. Slice leftover pork tenderloin for a sandwich and top with this for a lunch to kill.

1 small sprig each of any combination of:
   thyme, rosemary, lavender, parsley,
   or sage

½ cup (120 ml) water

½ cup (120 ml) **white vinegar**

2 teaspoons (7 g) **yellow mustard seeds**,
   toasted

¼ cup (50 g) sugar

¼ cup (85 g) honey

1 fresh jalapeño pepper, sliced in half,
   seeds and ribs removed

½ pound (225 g) **golden raisins**

Tie fresh herbs together with string or cheesecloth to form a sachet. Add it to a saucepan with the rest of the ingredients and bring to a quick boil. Reduce heat and simmer for 5 minutes until liquid reduces by half. Remove from heat and allow to come to room temperature. Transfer raisins and liquid to a tightly covered container and refrigerate until serving. Remove herb sachet and jalapeños before eating.

Yield: ¾ cup (195 g)

🍴 Cooked raisins evoke
the fall and accompany the
season's roasted meat dishes
especially well.

"LET'S TANGO, MANGO"

MANGO + CINNAMON + VINEGAR

Mangoes have an intensely sweet and luscious texture that is hard to resist—so don't! Cinnamon gives the mango a contrasting flavor component compatible with both savory and sweet applications. Adding vinegar tweaks the flavor just enough to elevate this combination to delicious heights.

The application recipe prepares a unique tangy mango ketchup. For another option with these components, purée them for a chilled soup (add additional juice to the application recipe). The flavors taste great with seafood and chicken, as well as fresh ginger and cardamom.

MANGO

Its juice is a thick mouthful. Use it fresh for clarity of flavor in salads and relishes. Cooking softens the texture and brings out its natural sugars. When dried, it becomes sweeter and its shelf life extends. For less than perfectly ripe mango, consider waiting a few days before using it or roasting it with orange juice to bring out the sugars.

CINNAMON

The application recipe uses ground cinnamon as seasoning and toasts it to accent its flavor. Use sticks rather than ground cinnamon to infuse the flavor into liquids with more subtlety and less visual impact. Grind your own cinnamon by chopping the sticks roughly with a knife and then grinding in to a spice grinder or cleaned-out coffee-bean grinder.

VINEGAR

Stick with light vinegars for this combination—Champagne, rice, or sherry—although a basic white vinegar works as well. Avoid red wine or balsamic vinegars here (they're great for other combinations), as their strong tastes and dark colors will overwhelm the combination.

"LET'S TANGO, MANGO" RECIPE

Mango Ketchup

Use this condiment the same way you would use tomato ketchup. It is especially good on crab cakes or spread on a BLT.

1 tablespoon (8 g) chili powder

1 teaspoon (2 g) **ground cinnamon**

½ teaspoon ground nutmeg

2 teaspoons (5 g) ground cumin

2 teaspoons (10 ml) canola oil

2 cloves garlic, minced

1 **mango**, peeled, flesh chopped

¼ cup (45 g) chopped tomato

1 tablespoon (15 ml) **rice vinegar**

¼ cup (60 ml) pineapple juice

Salt

In a large, dry, nonstick skillet, toast the chili powder, cinnamon, nutmeg, and cumin until fragrant and beginning to smoke. Add the canola oil and stir in the garlic, mango, and tomato. Cook for 5 minutes on low heat. Add the vinegar and pineapple juice. Let cool slightly and transfer to a blender. Blend until smooth. Add salt to taste. Keep covered in the refrigerator for up to 1 week.

Yield: 1 cup (240 g)

Mangoes are in season during the summer months, so take advantage of the ripe fruit then, which will have intense natural sweetness.

"ASIAN TRIO OF TEXTURE"

JICAMA + CABBAGE + SESAME

This combination plays on textures—the sweet crunch of jicama with the spicy bite of cabbage (either softened or crisp for an interesting vegetable mouthful. Bringing in sesame adds an Asian flair and a powerful flavor boost.

The application recipe prepares the ingredients in a quick pickle, finished with a light, flavorful sesame-lime dressing. Another option for the combination: softened cabbage and pork stew with sesame-jicama salad garnish. These flavors pair nicely with shellfish, pork, and Asian food.

JICAMA

Crunchy, sweet, and high in fiber, jicama is a healthy addition to many dishes. It pickles well because it absorbs the seasoning but still retains its crisp texture. Jicama root can be large, more than you'll likely need for a particular recipe, so substitute it for celery in salads or in vegetable crudités.

CABBAGE

Cabbage is most flavorful when raw in salads or a sandwich wrap. It is cooked to soften its flavor in soups and stews. All of these preparations work with this combination, although keep in mind jicama's crunchy texture. Let the overall texture of the dish guide how to handle the cabbage.

SESAME

Use the seeds to impart a nutty accent to dishes. White and black have a similar flavor, and combining them provides visual interest. Look to sesame oil to impart a rich sesame flavor, but use with a light hand as it's quite strong. Combine seeds and oil to give a dish a depth of sesame flavor.

"ASIAN TRIO OF TEXTURE" RECIPE

Pickled Vegetable Salad

This crunchy sweet-and-sour salad is great as a side dish or a topping for pork sandwiches, with seared ahi tuna, or with rice dishes. Slicing your vegetables on a Japanese mandoline is the best way to get thin, uniform "matchsticks."

2 cups (475 ml) water

²/₃ cup (160 ml) rice vinegar

1 cup (200 g) sugar

1 bay leaf

1 star anise

3 cloves garlic, peeled and sliced in half

½ inch (1¼ cm) fresh ginger, peeled and cut into 4 more pieces

1 fresh jalapeño pepper, stemmed and sliced in half

⅛ head **green cabbage**, shredded

⅛ head **red cabbage**, shredded

1 carrot, cut into matchsticks

½ red onion, sliced thinly

½ **jicama**, cut into matchsticks

2 scallions, sliced thinly into matchsticks

½ bunch fresh cilantro, leaves only, chopped

Juice from 1 lime

2 teaspoons (10 ml) **sesame oil**

Salt and pepper

In a large saucepan, combine water, vinegar, sugar, bay leaf, star anise, garlic, ginger, and jalapeño. Bring to a boil, stirring to dissolve the sugar. Add the cabbages, carrot, onion, and jicama. Cook for 3 minutes, until the carrot and onion are just tender. Drain liquid and remove bay leaf, star anise, jalapeño, garlic, and ginger.

In a large bowl, mix the cooked vegetables with the scallions, cilantro, lime juice, and sesame oil. Season with salt and pepper. Serve at room temperature or chilled.

Yield: 4 servings

When you have other vegetables on hand—such as snow pea pods or asparagus tips—feel free to add them into the mix.

"SOUR TURNED UPSIDE DOWN"

CABBAGE + DILL + CREAM

This combination captures the versatility of cabbage and combines it with dill's brightness and cream's richness. These flavors go especially well with chicken, seafood, beets, cucumbers, and sweet peppers. The application recipe uses cabbage in the form of sauerkraut and purées it for a lively and refreshing chilled soup with sweet crabmeat salad.

CABBAGE

Cabbage is most flavorful when raw; use it in salads, wraps, or for a slaw with buttermilk-dill dressing. Stuff cabbage leaves with wild rice and bake, serving them with a dill cream sauce. Or cook cabbage for a creamy soup with toasted dill croutons.

DILL

Dill weed is prized for its clean, pleasant flavor when fresh (don't bother with the too-mild dried dill). Dill weed (the frilly green ends) and dill seed (the small dried balls) differ pretty significantly in flavor, so note for which one a recipe calls. Fresh dill weed should not be cooked for an extended time because heat exposure diminishes its flavor, but toasting actually accentuates dill seed's flavor.

CREAM

Use heavy cream for the richest flavor. Buttermilk is also a nice choice for this combination because it offers flavor and creaminess without the calories or fat of heavy cream. Sour cream or crème fraîche are also good options for their tang (especially when using sauerkraut) and thickness.

"SOUR TURNED UPSIDE DOWN" RECIPE

Chilled Sauerkraut Soup with Crab Salad

This recipe, based on a Russian soup called chlodnik, makes a nice lunch option. A crab salad accents its smooth tanginess.

Soup

1 cucumber, peeled, seeded, and diced

1 teaspoon (6 g) salt

2 cups (475 ml) **buttermilk**

1½ cups (345 g) **crème fraîche**
(or sour cream)

½ cup (71g) **sauerkraut**, puréed in its juice in a food processor or blender

1 clove garlic, minced

2 tablespoons (8 g) chopped **fresh dill**

⅓ cup (28 g) finely chopped fennel bulb

1 scallion, finely chopped

Salt and pepper

Crab Salad

1 pound (455 g) lump crabmeat

Salt

½ cup (115 g) aioli or mayonnaise

To serve

4 sprigs fresh fennel

Fennel pollen, for garnish (optional)

To prepare soup: Place chopped cucumber in a colander and sprinkle with salt. Allow to drain onto a plate or in the sink for 20 minutes.

In a large bowl, whisk together buttermilk, crème fraîche (or sour cream, if using), puréed sauerkraut, garlic, dill, fennel, scallion, and salted cucumber. Season with salt and pepper. Chill for at least 4 hours or up to 1 day before serving.

To prepare crab salad: Place crabmeat in bowl and season lightly with salt. Mix with aioli and chill until serving.

To serve: Divide crab salad among 4 shallow bowls, creating a mound in the center of each. Ladle the chilled soup around the mound of crab salad. Garnish with a sprig of dill and a sprinkle of fennel pollen (if using). Serve immediately.

Yield: 4 servings

Although cabbage peaks in the fall, it is readily available year-round, especially high-quality purchased sauerkraut. This dish is quite refreshing during warm-weather months.

"RICH, SALTY...SUBSTANTIAL"

BEAN + BACON + VINEGAR

Their mild flavor and versatility make nutrient-rich beans a foundation for many dishes. They're also a great option for vegetarians—just not in this combination! Here, bacon makes the beans richer, saltier, and more substantial. Vinegar adds the sour edge that wakes up the combination's flavors. This trio makes a great side dish to take to a potluck or to serve on the picnic table with whatever is coming off of the grill.

BEAN

Check out the variety of dried beans in your grocery store: lima, black, pinto, cannellini, navy, cranberry—the list goes on and on. Use canned for convenience only. With dried beans, you can control the salt. Plus, they are cheaper and have a better texture. Cooked beans freeze well, so consider making a big batch and freezing some.

BACON

The crumbled bacon in this combination adds a meaty morsel and imparts nice flavor (the fat is used to sauté the vegetables). Layering the bacon flavor adds to the deep, satisfying richness in this dish.

VINEGAR

This dish needs vinegar to get its distinct sour flavor. Try cider vinegar for this combination. The apple tones of this strong vinegar add richness to the flavors. Sherry vinegar or a white vinegar also work for these ingredients.

"RICH, SALTY...SUBSTANTIAL" RECIPE

Kicked-Up Navy Beans

This recipe calls for dried beans. If you are not able to soak the beans overnight, cover them with water, bring to a boil, and let sit for an hour for a "quick soak." Of course, no one will report you to the authorities if you use canned.

1 pound (455 g) **dried navy beans**
　(or other bean of your choice)
4 slices **bacon**, cut into ¼-inch
　(²⁄₃ cm) pieces
1 onion, finely chopped
1 teaspoon (4 g) mustard
½ cup (115 g) packed brown sugar
¼ cup (60 ml) **cider vinegar**
1 tablespoon (18 g) salt, plus more
　for seasoning
1 cinnamon stick
Pepper

Wash the beans and soak overnight in water. The next day drain water, cover with fresh water, and cook over medium heat. Meanwhile, fry the bacon in a skillet until brown and crispy. Remove from skillet and add to the beans.

Add the onion to the skillet, keeping the bacon fat. Sauté until translucent, 2 to 3 minutes. Add the mustard, brown sugar, vinegar, 1 tablespoon (18 g) salt, and cinnamon stick. Stir and let simmer until the brown sugar dissolves.

Scrape contents of skillet into the pot of cooking beans and stir to combine. Let simmer 30 minutes to 1 hour until beans are tender. Season with salt and pepper. Remove cinnamon stick before serving.

Yield: 8 servings

Canned beans tend to have salt already, so be sure to taste your beans before adding any more. However, when you use dried beans, you may be surprised by just how much salt you need to add to get a dynamic flavor. Don't be afraid of the salt!

"CURRANTS GONE CREATIVE"

CURRANT + CHILE PEPPER + VINEGAR

This lively combination features the currant—a small, sweet grape most frequently seen dried but also available fresh. The heat of chile peppers tempers the sweetness that currant offers; the application recipe calls for jarred sweet cherry peppers for their sweet-hot flavor. Vinegar adds a sour element, giving this combination a vibrant, memorable flavor. Try these flavors as a sauce with chicken or pork, a relish for grilled sausages or hot dogs, or in a spicy jelly or jam.

CURRANT

Available in red, black, or white, currants have their fresh season during the summer months. Fresh, they are tart and juicy, similar to gooseberries; dried, they are raisin-like and sweet. To bring their flavor to the forefront, make currant jelly with a small amount of chile pepper for heat. Use raisins as a substitute, if necessary.

CHILE PEPPER

This combination takes advantage of jarred peppers, a quick way to harness the flavor of peppers without cooking. The application recipe uses sweet cherry peppers, also great stuffed with savory cheeses for a quick appetizer. Substitute whole peperoncini peppers or for a spicier outcome use jalapeños.

VINEGAR

For a less-sour taste, reduce the quantity of vinegar but keep a small amount in the mix for balance. Use a white vinegar, such as rice vinegar, to preserve the color of your sauce. A red vinegar gives a nice flavor but darkens the sauce in this application recipe.

Currant and Cherry Pepper Chicken

Currants sweeten this dish in two ways—as a jam, which enriches the sauce, and dried, cooked into the orzo.
If you have difficulty finding currant jam, try raspberry or apple.

1½ cups (355 ml) carrot juice
   (fresh or bottled)
2 tablespoons (30 ml) olive oil
1 onion, diced
1 celery stalk, diced
4 cloves garlic, thinly sliced
Zest of 1 orange
Juice and zest of 1 lime
2 chicken breast halves, bone in
   (about 12 ounces [340 g] each)
¼ cup (60 ml) **rice vinegar**
3 tablespoons (28 g) **dried currants**
1 tablespoon (20 g) **currant jam**
8 jarred **sweet cherry peppers**
1 teaspoon (5 ml) Sriracha sauce
½ cup (80 g) orzo (rice-shaped) pasta

Over medium-high heat, cook carrot juice in a saucepan for 10 minutes until reduced to ½ cup (120 ml). Set aside.

In a large pot or Dutch oven, heat oil over medium heat. Add onion and celery and cook until soft and translucent. Add garlic and cook for 2 more minutes. Add reduced carrot juice, orange zest, lime juice and zest, chicken, vinegar, currants, jam, cherry peppers, and Sriracha. Bring to a simmer, cover, and cook for 20 minutes.

Meanwhile, cook orzo in boiling, salted water for 8 to 10 minutes until al dente. Drain and reserve.

Cut chicken in half, checking to make sure that it is cooked thorough. Serve immediately with orzo and sauce.

Yield: 4 servings

Feel free to use any small pasta with this preparation—Israeli couscous, orecchiette, or farfalle all make nice options.

"SWEET TEA WITH A PUNCH"

TEA + LIME + SUGAR

Although these sound like beverage ingredients, they become the flavor accents for a cure for fish when used in a sweet-and-sour application. Sugar provides the sweetness, tea imparts an additional layer of flavor (varying depending on the type of tea), and lime gives both a sour element and a flavor punch. The type of tea used, as well as what is being cured, can influence the resulting flavor.

The application recipe uses salmon, but also try black cod or trout. Or take these same ingredients in a completely different direction in desserts such as granita, gelées, or as a poaching broth for fruit.

TEA
Earl Grey is a black tea flavored with bergamot (a small orange), which pairs nicely with the lime component. It is used in the application recipe, along with a bit of fresh orange zest to accent the citrus. To give the flavors a more intense tea taste, use Lapsang Souchong, smoke-dried over fires helping it retain a smoky flavor. For a mild subtle flavor, choose a green tea.

LIME
Its juice is the sour element for this cure, and the zest gives an additional boost of flavor. To highlight the lime element more, choose a green tea, which is milder. When buying a lime, choose one that seems heavy and yields under your thumb when pressed lightly; it will contain more juice.

SUGAR
For the sweet element in our combination, we use table sugar (or granulated). For liquid applications, consider using a simple syrup. Combine equal parts sugar and water, cook until dissolved and then refrigerate. It is useful for sweetening drinks of all kinds, as well as for chilled soups or vinaigrettes.

"SWEET TEA WITH A HARD RIGHT HOOK" RECIPE

Tea-and-Lime Cured Salmon

Although curing used to be necessary to preserve fish, in the age of refrigeration, it is continued for its ability to impart and intensify flavors. Be sure to start this preparation at least 24 hours before serving.

1¼ cups (250 g) **sugar**

½ cup (150 g) kosher salt

¼ cup (60 ml) brewed **Earl Grey tea**

Juice and zest from 2 limes

Zest from 1 orange

1 tablespoon (15 ml) dry sherry

1 teaspoon (2 g) C-Spice Blend (page 53)

1 large skinless, boneless salmon fillet (10 to 12 ounces
   [285 to 340 g])

Combine all of the ingredients except for the salmon in a large bowl. Mix, adding a little water, if necessary, to reach the consistency of a spreadable paste. Rub cure on all sides of the salmon fillet, turning it over and covering all visible flesh. Place on a tray and leave uncovered in the refrigerator for at least 24 hours. After 24 hours, check that the salmon is very tender; leave the cure on for additional hours if not.

When ready to serve, scrape the cure off of the salmon. Using a very sharp knife, cut diagonally into very thin slices. Try to make them as transparent as possible. Serve immediately or wrap and store for up to 1 week in the refrigerator.

Yield: 3 to 4 servings

This salmon makes a nice canapé, a topping for bagels and cream cheese, or served with eggs of all kinds.

"FULL-BODIED COMPLEXITY"

COFFEE + FIG + VINEGAR

This combination pursues the sweet, sour, and bitter. Coffee brings the rich bitterness, fig the luscious sweetness, and vinegar the sourness.

The application recipe prepares a modified gastrique sauce for pork tenderloin, which has an elusive and intriguing flavor. These flavors also are well suited to desserts: roasted figs over espresso custard and balsamic reduction or coffee ice cream with a fig-sherry vinegar sauce.

COFFEE

Espresso is the best choice for its richly concentrated flavor; French press or strongly brewed drip coffee works as well. Strong flavors are at work here, so you'll need strong coffee for the combination. There is really no place here for instant coffee.

FIG

Use the black-skinned, crimson-fleshed Black Mission fig in this combination. If you can't get your hands on fresh, dried works, too. Be sure to rehydrate them, in coffee if possible, before use.

VINEGAR

Adjusting which type of vinegar you use will allow for different results. There's a whole world of specialty vinegars—including fig vinegar, which would be a natural in this combination. Fruit-based vinegars, such as raspberry, could add an interesting element as well. For a lighter vinegar presence, use sherry vinegar; to deepen its contribution, use balsamic vinegar.

"FULL-BODIED COMPLEXITY" RECIPE

Pork with Sweet and Sour Citrus Sauce

This recipe uses C-Spice Blend (page 53) to amplify and deepen the flavors. If you don't have it or cannot make it, double the amount of the sugar-cocoa spice for the figs, reserve half, and add salt and pepper. Use this as a rub for the pork tenderloin.

2 ¼ pounds (1 kg) **fresh figs**
    (preferably Black Mission), divided

⅓ cup (67 g), plus 2 teaspoons
    (8 g) sugar, divided

2 tablespoons (12 g) C-Spice Blend
    (page 53), divided

1 teaspoon (2 g) unsweetened
    cocoa powder

¼ cup (60 ml) **balsamic vinegar**

1 cup (235 ml) brewed **espresso**

Zest of ½ orange

Juice from ½ lime

Salt

2 whole pork tenderloins

This dish is well suited to the fall months, when figs are in season and the richness of pork sounds appealing as cool weather approaches. Also, try the sauce with duck or tuna for a nice variation.

Preheat oven to 325°F (170°C, or gas mark 3). Choose 7 figs to roast and set the others aside. Slice the 7 figs in half lengthwise and place cut-side up on a sheet pan. In a small bowl, combine 2 teaspoons (8 g) sugar, 1 teaspoon (2 g) C-Spice Blend, and cocoa. Sprinkle liberally over the figs and bake for 10 minutes or until tender but not mushy. Cool, dice into small pieces, and set aside.

In a large, heavy saucepan, cook the ⅓ cup (67 g) sugar over medium-high heat to a dark caramel. Add the vinegar, let it bubble and splatter, and then add the espresso, orange zest, and remaining figs. Cook until the figs are quite soft.

Let cool slightly and purée mixture in a blender. Pass through a fine mesh strainer, add lime juice, and adjust seasoning with salt. Add roasted chopped figs to smooth sauce and place in a saucepan until pork is ready.

Trim any visible silver skin from the pork tenderloins. Rub all over with remaining C-Spice Blend and sear all sides in a large skillet until browned. Cook in the oven for 10 minutes or until the internal temperature reaches 140°F (60°C). Let pork rest for 5 to 10 minutes before slicing. Warm sauce and serve on top.

Yield: 4 servings

CHAPTER 6

Exotic Flavors

It's a big world out there, with many rich culinary traditions to offer inspiration. And though world travel is a luxury unavailable to many of us, cooking from an ethnic perspective can give us a bit of armchair travel, exposing us to the flavors of faraway lands from the comfort of our own kitchens.

Part of the fun of cooking recipes inspired by foreign cuisines is the pursuit of unfamiliar ingredients. Just as surfing the Web about your next vacation can extend the pleasure of the actual trip, searching for an exotic flavor can amplify enjoyment of the dish. Never heard of bonito flakes? By the time you drive across town to an Asian market you have never visited, exposing yourself to aromas and unknown languages, you may more appreciate the resultsof preparing your own dashi. In the era of the Internet, even geographic isolation cannot exclude you from acquiring virtually any ingredient.

The recipes in this chapter represent different world cuisines—Asian, Middle Eastern, Latin American, European, and Indian. This is certainly not an exhaustive list, and the recipes aren't necessarily designed to be authentic representations of an area's cuisine. Instead they represent flavor combinations typical to various parts of the world and cooking techniques readily applied.

For example, Asian recipes commonly include soy, miso, ginger, or sesame, in a range of dishes from vegetables to seafood. Similarly, the reliance of Middle Eastern cuisine on nuts, fresh herbs, legumes, and cardamom provides a profile as distinct as it is flavorful. The spiciness and prominence of chile peppers and cilantro in Latin American food translate into unique salsas, sauces, and main courses.

And in case you think European influences lack the pizzazz of other parts of the world, remember that butter, olive oil, and garlic have their roots there and can improve the flavor of virtually any ingredient. Finally, look to Indian cuisine and its wide range of curries to enliven many dishes, creating exotic tastes.

Once you get the hang of creating an ethnically inspired taste profile—as in, "I think I'll add a bit fresh ginger, sesame oil, and a splash of soy sauce to this salad dressing"—nothing can stop you. The world is actually at your fingertips, in the form of food. You may not get a vacation for another six months, but you can travel in your kitchen. Enjoy finding bottarga, sumac, or chayote to cook with and take off!

"NUTS GONE EAST"

PEANUT + SOY SAUCE + CHILE PEPPER

Peanuts bring flavor that transcends type—from a sandwich spread and salty snack to chocolate cup or sweet brittle, they always please. Peanuts are taken to the Asian persuasion in this combination. Salty soy sauce lightens their substantial peanut taste, and spicy chiles hold their own with the rich, nutty flavor, giving the combination a vibrant heat. These flavors work well with tofu, chicken and pork, or rice dishes. Alternative applications include a spicy dip for pita bread or vegetarian chili.

PEANUT

Roasting the nut intensifies its flavor. It works well in savory and sweet dishes, as well as puréed into butters. Its oil has as mild flavor and high burning temperature—characteristics many cooks appreciate. Peanuts are less expensive than many other nuts, so take advantage of their crunchy, full flavor.

SOY SAUCE

Made from fermented soybeans and grain, soy sauce has the distinct taste called umami by the Japanese—translated to "delicious taste"—and not one of the other four basic tastes (sweet, sour, salty, and bitter). It is used primarily as a condiment or as a flavor component in sauces or vinaigrettes.

CHILE PEPPER

Fresh chiles are the most potent when eaten raw but soften considerably when roasted. Dried chiles are readily available and are rehydrated upon use. For flavorful seasoning, use powders or liquid sauces. For this combination, Thai chiles, fresh or dried, go perfectly with the bold, strong flavors.

"NUTS GONE EAST" RECIPE

Asian Peanut Sauce

This peanut sauce starts with uncooked peanuts for the purest taste possible. In a pinch, you can use one cup (260 g) of natural unsalted peanut butter. Serve over tofu, chicken, pork, rice, or noodle dishes.

1 cup (145 g) **peanuts**, uncooked and unsalted

1 teaspoon (5 ml) sesame oil

¼ teaspoon salt

1 tablespoon (6 g) chopped fresh ginger

2 cloves garlic, chopped

½ cup (100 g) sugar

¼ cup (60 ml) mirin (cooking sake)

¼ cup (60 ml) rice vinegar

½ cup (120 ml) **soy sauce**

2 tablespoons (30 ml) **Sriracha sauce**

4 sprigs fresh cilantro (optional)

In a skillet over medium heat, roast peanuts in sesame oil and salt until golden brown. Add all remaining ingredients except for the cilantro and cook together for 3 minutes. Allow to cool for 10 minutes.

Add peanut mix and cilantro (if using) in blender and process until smooth. Make this sauce up to 2 days prior to serving and reheat before using.

Yield: 1 ½ cups (390 g)

Feel free to increase the chile component if you like a little extra heat.

"TURNIPS IN A KIMONO"

TURNIP + MISO + MIRIN

Turnips don't often star in a show, but hide with the orchestra, as an ingredient in a winter soup or mashed with potato, for example. Here they belt out a solo—accented by the Asian-influenced understudies of miso and mirin. The former brings an earthiness to the turnips, and the latter accents their sweetness while balancing the flavors.

These flavors together in the application recipe create a simple, flavorful side dish. As an alternative, slice them thinly for a salad with miso-mirin vinaigrette. Think of this combination for pork, rice, and tofu preparations.

TURNIP

This root vegetable, in season during the summer months, has a radish-like flavor. Look for the smallest turnips you can find (baby or Japanese turnips), considerably sweeter and more tender than the large variety and can be eaten raw in salads. Save the greens for cooking—serve with the root for added turnip flavor or cook alone for a taste similar to mustard greens. Preserved turnips (sold in Asian markets) add another sweet and tangy layer to any dish and keep within the Asian flavor profile.

MISO

If you have a choice, go for an unpasteurized miso for its improved taste and health benefits. Choose a white miso for the most versatility, as it has the mildest flavor. For a tasty alternative, turn this combination into a simple miso soup with diced turnips and a mirin accent.

MIRIN

This Japanese cooking wine made from rice sweetens and flavors vegetables, seafood, sauces, and glazes. As a substitute, combine sake with a bit of honey.

"TURNIPS IN A KIMONO" RECIPE

Turnips with Mirin Sauce and Miso Butter

A miso butter accents this dish as a finish. Make extra of this flavorful butter and use it to season any type of cooked vegetable.

3 tablespoons (48 g) **miso paste**

¼ cup (55 g) unsalted butter, softened and divided

3 pounds (1.4 kg) **small turnips**, halved

1½ cups (355 ml) water

½ teaspoon salt

1 tablespoon (15 g) packed brown sugar

2 tablespoons (30 ml) **mirin** (cooking sake)

Greens from turnips, stems discarded and leaves chopped

1 tablespoon (15 ml) yuzu juice (or lemon juice)

Mix miso paste and 3 tablespoons (42 g) of butter; set aside.

Place turnips in a large saucepan and cover with water, salt, remaining butter, brown sugar, and mirin. Bring to a boil, cover, and cook for 8 minutes.

Add greens and stir to mix. Add yuzu (or lemon) juice and cook for an additional 1 minute. Add miso butter and cook until the liquid reduces and the turnips are tender. Serve immediately.

Yield: 4 servings

 Turnips are a summer vegetable, often showing up at farmers' markets in the small, colored varieties— yellow, orange, and red, in addition to white. They are perfect for this recipe and for salads.

"BANGKOK DELIGHT"

LEMONGRASS + COCONUT + BASIL

The natural sweetness of coconut makes a lovely base for the subtleties of lemongrass and the herbal sweetness of fresh basil. Southeast Asian cuisine typically combines these ingredients for use as a versatile sauce or soup with many applications. The flavors work well with shellfish, fish, chicken, tofu, and curry. Add them to rice or noodles to make them more substantial. Or combine the ingredients for sweet applications—coconut cupcakes with lemongrass-basil–infused buttercream or coconut-lemongrass tapioca with citrus-basil sorbet.

LEMONGRASS

Lemongrass looks nothing like lemon, yet it imparts a lemony flavor and intense aroma to dishes without lemon's acidity. To let loose the flavor of the tough, slender stalk, cut it or "bruise" it (put firm pressure on it using the back of a knife). If purchasing dried, soak it in hot water for at least 30 minutes before using. If given a choice, stick with fresh.

COCONUT

Coconut milk, sold in cans in most grocery stores, is a traditional base for Thai cooking. It is made when coconut juice and meat are puréed and strained, creating a rich liquid. Because its flavor is relatively mild, use it liberally with this combination. Also look for coconut juice (sometimes called coconut water), the liquid from inside the coconut that has a delicate, sweet flavor and a lighter consistency than the milk. You may use the two interchangeably, but the juice results in overall lighter flavors.

BASIL

Avoid dried basil, which loses most of its flavor and is weaker than fresh basil. For this combination, use the assertive Thai variety for its slight spiciness. Thai basil, widely used in Southeast Asian cuisine, can be found in most Asian markets.

"BANGKOK DELIGHT" RECIPE

Heavenly Thai-Inspired Mussels

The broth in this dish is like liquid heaven, so make sure you have something on hand to help you gobble it up—rice, bread, or even just a spoon will do.

1 can (14 fluid ounces, or 425 ml)
  coconut milk

½ inch (1¼ cm) fresh ginger, peeled
  and minced

2 **stalks lemongrass**, trimmed and
  chopped into 2-inch (5 cm) pieces

1 fresh Thai chile, sliced lengthwise,
  seeds removed, divided

Zest of 1 lemon

1 tablespoon (15 ml) fish sauce
  (optional)

Juice of 1 lemon

2 cloves garlic, minced

1 small shallot, minced

2 tablespoons (30 ml) vegetable oil
  (such as canola)

⅛ teaspoon sesame oil

¼ cup (60 ml) dry white wine
  (such as Sauvignon Blanc)

1 pound (455 g) mussels, washed

6 **stems Thai basil**, leaves only,
  chopped

Salt

In a saucepan, combine coconut milk, ginger, lemongrass, half of the chile, lemon zest, and fish sauce (if using). Bring to a simmer and cook for 10 minutes. Remove from heat, cover, and let infuse for an additional 10 minutes. Strain, pushing down on the strainer to extract as much liquid as possible. Add lemon juice and set aside.

Finely mince the remaining half of the chile. In a large skillet over medium-high heat, sauté chile, garlic, and shallot in vegetable and sesame oil until tender. Deglaze pan with white wine and cook until reduced by half. Increase the heat to high; add mussels, coconut-lemongrass milk, and basil. Season with salt and shake the pan to distribute the ingredients. Cook, shaking the skillet occasionally, just until the mussels open. Remove from heat and serve immediately.

Yield: 4 servings

This recipe also works nicely with shrimp as a substitute for the mussels.

"ROCKIN' MOROCCAN"

CILANTRO + CUMIN + TURMERIC

This combination captures the robust flavors of cumin and turmeric and balances them with the freshness of cilantro for a delightful flavor. Toasting the cumin seeds will deepen the seeds' flavor, but the components do not need to be cooked to make them enjoyable. The trio is a staple in Middle Eastern cuisine, and the application recipe is based on a Moroccan chermoula, a paste of fresh and dried spices used as a marinade for fish or chicken or as a seasoning for rice and vegetables.

CILANTRO

Because cilantro's flavor decreases quickly once it is dried or frozen, it is best to use the herb fresh. To highlight its flavor, add it toward the end of any cooking process; otherwise, prolonged heat dulls its impact.

CUMIN

Buying whole seeds, toasting them, and grinding them yourself provides the most flavor. If buying ground cumin, purchase small amounts at a time, as it (like most ground spices) quickly loses flavor over time.

TURMERIC

This bright yellow spice most commonly acts as a base for curries, but it also seasons meat and vegetables in Middle Eastern cuisine. Use it as an accent, but resist the urge to let it dominate—its powerful color can create an artificial appearance.

"ROCKIN' MOROCCAN" RECIPE

Robust Chermoula

Use this versatile, flavorful combination with seafood, chicken, rice, or vegetable dishes. Try sliding tilapia into it before broiling or toss roasted cauliflower in it.

2 cups (32 g) **fresh cilantro** leaves

1½ cups (90 g) fresh Italian parsley leaves

4 cloves garlic, peeled

½ teaspoon salt

2 teaspoons (4 g) **cumin seeds**, toasted

2 teaspoons (4 g) coriander seeds, toasted

2 teaspoons (4 g) **ground turmeric**

2 teaspoons (5 g) sweet paprika

½ teaspoon hot pepper flakes

⅛ teaspoon cayenne chili powder

¼ cup (60 ml) lemon juice

⅓ cup (80 ml) olive oil

Add all the ingredients except for the oil to a food processor. Pulse several times to chop together until ground. With the motor running, add the olive oil in a steady stream, forming a paste. Add additional oil if using for a marinade. Keep covered in the refrigerator for several days.

Yield: 1 cup

Because of the vibrant flavor of the fresh herbs, plan on preparing this no more than a day before you intend to use it, as the color and flavor will diminish quickly.

"WORLD HARMONY"
PLANTAIN + CURRY + VANILLA

Plantains—not to be confused with bananas—are a staple in African and Caribbean cuisine. They are starchy and low in sugar and must be cooked before being eaten. Their mild flavor can handle a little spice, so curry keeps the ethnic flavor relevant. Introduce vanilla and things get interesting—its heady aroma is not sweet, yet it adds a nuance to the combination that brings out the plantain's sugars and deepens the curry notes.

The application recipe uses the flavors to produce spiced plantain chips. For another option, fry the ingredients into a fritter or slow cook them as a stew to serve with rice.

PLANTAIN
Used more like a potato than a fruit in cooking applications, plantains are available at three stages of maturity: green (bland and hard), yellow (ripened and slightly sweet), and black (completely ripe and very sweet and aromatic). Note for which type your recipe calls; it will greatly impact the dish.

CURRY
In India, the flavors that compose curry can be quite diverse, but in Western cuisine, it has become a generic description of a spice blend. The spices generally include turmeric (giving it the characteristic yellow color), cumin, coriander, and fenugreek (used for its tang and intensity), among others. The powder form (also called garam masala) is most easily obtained; purchase it in small quantities, as its flavor quickly fades.

VANILLA
Do not substitute vanilla extract for vanilla beans in this combination. In addition to the complementary aromatic qualities it offers, vanilla's visual aspect—its little black specks—is important. They add a level of intrigue and surprise in a savory dish, and reveal the use of high-quality ingredients. This recipe is the perfect opportunity to reuse a vanilla bean you have saved from a previous application. If starting with a soft, fresh bean, dry it out in a low-temperature oven until brittle enough to grind.

"WORLD HARMONY" RECIPE

Crispy Plantain Chips with Curry-Vanilla Sugar

A mandoline slicer greatly simplifies preparation of these spicy chips, providing very thin slices and adding crispiness. It is also less risky—you will be less prone to cutting yourself than with a very sharp knife. Try a vegetable peeler for a similar result.

2 green **plantains**, peeled and sliced
   as thinly as possible

1 tablespoon (6.3 g) **curry powder**

½ dried **vanilla bean**

1 teaspoon (4 g) sugar

1½ teaspoons (9 g) salt

1 cup (225 g) unsalted butter (2 sticks)

2 cups (475 ml) canola oil

Place sliced plantains in a bowl of ice water to reduce the starchiness in the chips. Meanwhile, in a spice grinder (or a cleaned-out coffee-bean grinder), combine curry powder, vanilla bean, sugar, and salt until the mixture contains no large vanilla pieces.

Spread out plantain pieces on paper towels to drain and pat dry with a clean paper towel. In a deep pan, heat butter and oil over medium-high heat until the oil reaches 350°F (180°C).

Lower a handful of plantains into the oil with tongs and cook until browned all over, stirring to separate the chips as they cook. Drain on paper towels and sprinkle with curry-vanilla powder while still hot. Repeat with remaining plantains and serve immediately.

Yield: 4 cups chips

These chips make a unique snack or appetizer with a ginger beer—a great start to a meal of pork loin and black beans.

"SMOKY INTRIGUE"

DATE + CHESTNUT + PAPRIKA

This combination has a bit of intrigue. The ingredients are familiar, yet they have a whiff of the exotic. The sweet date joins the delicate nuttiness of chestnut and gets a jolt of sweet smokiness with paprika.

The application recipe prepares an unusual dish accented by a toasted pasta and smoked olive oil. Also look to the combination for stuffed dates as an appetizer or as a stuffing with game birds. These flavors complement pork, lamb, sweet potatoes, mushrooms, and carrots.

DATE

Because of their high sugar content, dates can be used easily in desserts but also as sweet accents to salads, curries, and couscous. Look for soft dates with a seemingly liquid flesh contained by a thin, dried shell. For a natural sweetener (or to use dried-out dates) make date "sugar." Cook the fruit in the oven until they are rock hard (20 minutes at 350°F [180°C, or gas mark 4]) and then grind in a food processor.

CHESTNUT

Roasting gives a texture similar to baked potatoes, with a sweet, soft, delicate flavor. Other techniques that work effectively: Boiling, steaming, grilling, and frying. Canned are fine for purées, but rely on fresh in the shell for other applications. Removing the shells can be a bit of a chore; be sure to cut an X on the flat side before cooking or you risk an explosion! December is the prime month for fresh chestnuts—hence the Christmas song reference.

PAPRIKA

Paprika can range from slightly sweet to quite spicy. Look for sweet smoked paprika (or Pimentón de la Vera, Dulce). This Spanish paprika has a particularly sweet, cool, smoky flavor. It is a great way to add smoky flavor without heat, which could overwhelm the sweet subtleties of the chestnut. Regular paprika will not have the same vibrant flavor but can be used in this combination.

Sardinian Toasted Pasta with Chestnut Sauce

This recipe greatly benefits from use of a specialty pasta called Fregola Sarda, a Sardinian toasted durum wheat pasta with a unique, deepened flavor. Also try to get your hands on another specialty ingredient—smoked extra-virgin olive oil. For substitutes, use Israeli couscous and a high-quality extra-virgin olive oil.

1 tablespoon (7 g) **sweet paprika**

1 tablespoon (14 g) unsalted butter

½ yellow onion, diced

4 cloves garlic, diced

1 cup (235 ml) carrot juice

1 cup (235 ml) dry white wine
　(such as Chardonnay)

1 cup (235 ml) water

1 sprig thyme

3 bay leaves

8 large **dates**, pitted and diced

1 package (12 ounces, or 340 g) small
　pasta (ideally Fregola Sarda)

8 ounces (227 g) **chestnuts**, chopped
　(ideally fresh and roasted, or canned),
　with some larger and smaller pieces

Salt and pepper

1 tablespoon (3 g) finely chopped chives

2 tablespoons (30 ml) smoked olive oil

In a small, nonstick skillet over medium heat, toast the paprika until it begins to brown around the edges and starts to smoke. Remove from heat.

In a large saucepan, melt the butter over medium heat and add the paprika, onion, and garlic. Cook until the onion is translucent. Add the carrot juice, wine, water, thyme, bay leaves, and dates and bring to a boil. Add the pasta and cook for 10 minutes until most of the liquid absorbs and the pasta is al dente

Set aside the larger pieces of chestnut and fold the smaller pieces into the pasta. Season with salt and pepper. Remove the bay leaves and sprinkle with the large pieces of chestnut and chive. Drizzle with smoked olive oil immediately before serving.

Yield: 4 servings

🍴 This can be a stand-alone vegetarian dish great with a salad and baguette. Or serve it with a lamb or pork chop.

"EDGY ORANGE"

ORANGE + CARDAMOM + COCONUT

Cardamom provides an interesting contrasting flavor profile that takes some of the acid edge off of the citrus. Look for cardamom pods for maximum flavor and use the ground only if you can't locate the pods. Coconut with orange can take on a tropical flavor profile, but the cardamom pushes a more complex element. Take these flavors to the savory side by letting the ingredients speak for themselves, with little cooking needed. To sweeten them up for a dessert application, consider adding sugar to the coconut (for a custard, perhaps) and serving it with an orange-cardamom sorbet. These flavors also work nicely in beverages, with orange and coconut juices as the base.

ORANGE

Juice oranges for drinks and reductions, use their segments in salads and savory applications, preserve them for long storage, utilize the zest for seasoning and in cocktails, or candy the rind. The orange possibilities are limitless, especially with the other ingredients in this combination!

CARDAMOM

Sold as a whole pod or ground, this intensely aromatic seasoning is common in Middle Eastern, Indian, and Scandinavian cuisines. It is used in curries, rice dishes, sweet pastries, coffee, and desserts. Go easy with it in this combination so that the subtleties of the other flavors don't get lost.

COCONUT

The juice can be sipped directly from the coconut, and the flesh can be processed fresh or purchased flaked (sweetened or unsweetened). Fresh coconut is not as sweet as the processed variety. Coconut milk, sold in cans in most grocery stores, is a traditional base for Thai cooking. Because its flavor is relatively mild, use it liberally with this combination.

"EDGY ORANGE" RECIPE

Warmly Spiced Couscous Salad

Combining these flavors with couscous makes a great side dish or accompaniment for shrimp or a white fish such as sea bass.

2 cups (475 ml) water

3 **cardamom pods**

2 star anise

¼ cup (60 ml) medium dry-to-sweet wine
   (such as Riesling or Gewürztraminer)

2 tablespoons (30 ml) rice vinegar

2 tablespoons (30 ml) almond oil

Salt

1½ cups (265 g) couscous

1 can (8 fluid ounces, or 235 ml)
   coconut milk

¼ cup (25 g) diced green onions,
   green only

2 **oranges or tangerines**, divided into
   segments

½ cup (35 g) shredded Napa cabbage

½ cup (68 g) chopped macadamia nuts

Pepper

In a medium-size saucepan, bring the water to a boil with cardamom pods, star anise, wine, vinegar, almond oil, and a pinch of salt. Slowly add couscous to the boiling water. Cook, stirring occasionally, for 5 minutes or until liquid absorbs and the couscous is soft.

Remove from heat, stir in the coconut milk, green onions, orange or tangerine segments, Napa cabbage, and macadamia nuts. Season with salt and pepper. Serve at room temperature or cold.

Yield: 6 to 8 servings

This refreshing, interesting flavor combination tastes especially good in the warm summer months. However, with citrus at its peak during the winter, take full advantage of using it year-round.

"ARABIAN DELIGHT"

CHICKPEA + EDAMAME + CORIANDER

Chickpeas, otherwise known as garbanzo beans, have a mild flavor but a full and rich consistency. They make a substantial base for a dish, and when combined with the toothsome bite of edamame (a soybean), work well with a strong flavor component. Coriander provides that, taking the combination toward a Middle Eastern flavor profile.

This application purées the ingredients to form a tasty falafel patty. For an easier preparation, leave the ingredients whole, season them, and toss with pasta or salad greens. Or make a lively dip with the same components by puréeing them until very smooth. Serve with wedges of pita bread.

CHICKPEA

When using dried chickpeas, cook them to get a soft texture. Using canned products does not change the flavor significantly. Chickpeas are often combined with other beans or puréed into dips and can be fried to make a crispy garnish. In this combination, chickpeas provide the substantial base to incorporate the dish's flavors.

EDAMAME

These baby soybeans are either left in the pod or shelled and frozen. In their most simple form, serve them with a sprinkling of salt, either hot or cold. Combining them to chickpeas adds a nice green color and a new, fresh dimension for the falafel.

CORIANDER

The coriander seed has a nutty flavor accentuated by roasting. Buying whole seeds, roasting them, and grinding them gets the maximum flavor possible. Ground coriander is acceptable; just be sure to use it soon after buying, as its flavor diminishes over time.

"ARABIAN DELIGHT" RECIPE

Edamame Falafel

Make this version of falafel ahead of time. Just freeze the patties and pull them out when needed. Great toppings include Turmeric-Pickled Cauliflower (page 157) and tzatziki sauce. Tuck them all into pita bread for a filling, tasty sandwich.

½ yellow onion, diced

4 cloves garlic, diced

1 tablespoon (15 ml) olive oil, plus more to fry patties, divided

1½ teaspoons (3 g) ground **coriander**

1½ teaspoons (4 g) ground cumin

Pinch ground cardamom

¼ cup (7 g) chopped cilantro

1 cup (118 g) **shelled edamame**, thawed

1 can (15 ounces, or 430 g) **chickpeas**, rinsed and drained

Zest of 1 lemon

Juice of 1 lemon

½ cup (63 g) all-purpose flour

Salt

Sauté onion and garlic over medium heat in olive oil until translucent. Add them, along with all remaining ingredients except for the flour, to a food processor. Mix until combined and beans are uniform size, scraping the food processor bowl regularly.

Transfer mixture to a large bowl and fold in the flour. If the mixture does not hold together, add more flour 1 tablespoon (7.5 g) at a time.

Form mixture into 4 patties. Deep-fry or sauté them in a skillet until browned on each side. Serve immediately.

Yield: 4 patties

This is a great vegetarian offering that meat eaters will love as well. It is also a nutritious selection—high in fiber and protein, and low in fat.

"MIDDLE EASTERN MEDLEY"

SUMAC + SESAME + THYME

This combination of dynamic tastes reflects Middle Eastern flavors. The tart sumac flavor elevates the nutty sesame, and thyme's herbal mildness balances the mix.

The application recipe prepares a spiced blend called za'atar for use with a roasted chicken. Try the za'atar mixed with olive oil for bread or sprinkled on hummus, potatoes, or legumes. These flavors also combine well with beets and yogurt.

SUMAC

Middle Eastern cooking frequently calls for this dark-crimson powder. Its fruity, sometimes astringent flavor is available either as whole berries or ground. Use it whole in marinades and dressings and ground as a spice rub for meats or on vegetables. If you cannot locate it, substitute lemon zest—just know it won't produce quite the same taste.

SESAME

Use the seeds to impart a nutty accent to dishes. White and black have a similar flavor, and combining them provides visual interest. Look to sesame oil to impart a rich sesame flavor, but use with a light hand as it's quite strong. Combine the seeds and oil to give a dish a depth of sesame flavor.

THYME

This easily grown herb has a mild flavor that blends well with other herbs and ingredients. Garden thyme is the most readily available variety, but look for lemon thyme as well, prized for its lemon aroma. Though you can use dried thyme for this combination, it will lack fresh thyme's vibrant flavor.

"MIDDLE EASTERN MEDLEY" RECIPE

Za'atar Roasted Chicken

This flavorful spice mix is used as a rub for a roasted chicken.
Try the same technique with fish such as salmon.

2 tablespoons (5 g) chopped **fresh thyme leaves**

2 tablespoons (16 g) **sesame seeds**, toasted

2 teaspoons (5 g) **ground sumac**

½ teaspoon sea salt, plus more to sprinkle on lemons

2 lemons, quartered

1 whole chicken (2 to 3 pounds [900 g to 1.3 kg])

Olive oil

Preheat oven to 350°F (180°C, or gas mark 4). Combine the thyme, sesame seeds, sumac, and ½ teaspoon salt. Sprinkle lemons with salt and stuff inside the chicken cavity. Cover chicken with the spice blend and place in a roasting pan. Drizzle with olive oil.

Bake uncovered for 50 to 60 minutes or until the skin is crispy and the leg meat is no longer pink when cut into (about 160°F [70°C] internal temperature). Serve immediately.

Yield: 4 servings

Spice mixes such as this can add an exotic flair to your everyday cooking without much effort.

"PEPITAS ON A MISSION"

PUMPKIN SEED + CHILE PEPPER + GARLIC

Mexican cuisine inspired this combination. Pumpkin seeds, also known as pepitas, are crunchy morsels with a powerful flavor when combined with the spice of chiles and the boldness of garlic. Vary the intensity of the combination primarily by changing your use of the chile and garlic—raw is potent; sautéed or roasted take off some of the edge.

The application recipe prepares a sophisticated salsa, but also look to the combination for sauces such as mole or for enchiladas.

PUMPKIN SEED

Of course you can get them directly out of a pumpkin (which, if you aren't carving for Halloween, you should roast to serve with the salsa). The seeds also are widely available packaged in grocery stores. Look for the hulled variety (they should be dark green) for the most versatility. Roast or toast for deepened flavor and crunch, especially when using them for a snack or garnish. Grind or purée them for a sauce thickener.

CHILE PEPPER

Often associated only with heat, chiles offer so much more—flavor! The application recipe calls for güero peppers (sometimes called wax or banana peppers), pale, greenish peppers that are sharp when raw, but when roasted, gain a rich flavor milder than the jalapeño. For a spicier option, substitute jalapeños.

GARLIC

When using garlic raw, make it a practice to slice the peeled clove in half lengthwise to remove the greenish shoot that sometimes grows through the middle. When raw, these shoots can add an off flavor. Look for the red-skinned, small-cloved garlic for Mexican food. Available in the summertime, it offers a pungently sweet taste harmonious with these flavors.

"PEPITAS ON A MISSION" RECIPE

Pumpkin Seed Salsa

*Think beyond tortilla chips to highlight these wonderful flavors. Try on roasted pumpkin or other winter squash
as a vegetarian main course or side dish, on greens for a salad with fresh chevre, or on seared scallops.*

3 **fresh güero chiles**, seeded and
   sliced in half

1 medium white onion, peeled
   and quartered

1 large tomato, quartered

10 cloves **garlic**, peeled

3 tablespoons (45 ml) olive oil, divided

¼ cup (16 g) **hulled pumpkin seeds**

1 teaspoon (3 g) ground cumin

Juice from 1 lime

1 tablespoon (15 ml) sherry vinegar

¼ teaspoon Mexican oregano, toasted

½ cup (120 ml) water

Salt

Preheat the oven to 350°F (180°C, or gas mark 4). In a bowl,
toss the chiles, onion, tomato, and garlic with 2 tablespoons (30 ml)
of olive oil and 2 large pinches of salt. Spread on a sheet pan in a
single layer and cook until the skins of the peppers puff and blister,
15 to 20 minutes.

While the vegetables are in the oven, toast the pumpkin seeds. Heat
the remaining oil with the ground cumin; add the seeds and cook over
medium heat until browned and fragrant. Allow to cool and then chop
roughly by hand or in a food processor.

Remove the skins from the peppers. Add cooked vegetables to a food
processor with the lime juice, vinegar, oregano, and water. Pulse a few
times until you reach a chunky consistency. Transfer to a bowl and stir
in pumpkin seeds. Salt to taste. This salsa tastes best the day it is
made.

Yield: 2 cups (500 g)

🔧 Use this salsa as your
secret weapon to elevate
many dishes—even non-
Mexican ones. Make it
year-round but especially
in the summer when ripe,
home-grown tomatoes are
plentiful.

"LATIN AMERICAN FLAIR"
CACTUS + CHAYOTE + APPLE

This combination takes advantage of some underutilized ingredients found more extensively in Latin American cuisine. The unique flavor and texture of cactus (also known as nopales) and the mild crispness of chayote come together with the familiar taste of sweet apple to create a dynamic flavor combination.

The application recipe prepares a savory-sweet salad. You can also try the combination as a salsa or in a soup. These flavors interact nicely with lime, chile pepper, and citrus.

CACTUS
Don't let the preparation of cactus intimidate you or be an obstacle in your cooking. You must remove the thorns and spines, but you can usually accomplish this with a vegetable peeler or knife. It's no more time consuming than peeling potatoes. Serve cactus raw, grilled, or boiled—but be careful not to overcook it, which often creates a slimy texture. The application recipe uses the paddles, but also look for the fruit of the plant, called cactus pear, which has a sweet and sour flavor.

CHAYOTE
Handle this gourd, also known as Mexican squash, like a summer squash—serve it raw or lightly cook it. Its tubers are also edible, often used in the same way as other root vegetables such as potatoes. Chayote is available widely in Latin American markets or in the ethnic produce area of many grocery stores.

APPLE
Use raw apples to get the greatest texture and crisp flavor. For raw applications in this combination, seek out Pink Lady or Gala apples. Cooking softens the intensity of the apple's flavor and texture. Tart Granny Smith apples balance the sugars that increase as the apples cook. To let the apple flavor dominate, use it raw for crunch and prepare vinaigrette out of juice or cider to layer the apple flavor.

"LATIN AMERICAN FLAIR" RECIPE

South-of-the-Border Salad

Serve this flavorful salad with marinated grilled chicken or toss with chopped romaine for a heartier dish.

1 teaspoon (2 g) cumin seeds

1 teaspoon (2 g) coriander seeds

½ cinnamon stick

1 teaspoon (3 g) garlic powder

½ dried chipotle pepper, chopped

1 teaspoon (6 g) salt, plus more for
  seasoning, divided

½ pound (227 g) **fresh nopales**
  (3 prickly pear paddles), thorns
  removed with vegetable peeler, rinsed,
  and cut into ¼-inch (⅔ cm) slices

½ pound (227 g) **chayote** (about
  1 gourd), peeled, halved lengthwise,
  seeds scooped, and cut into 3-inch
  (8 cm) matchsticks

1 **apple** (preferably Granny Smith),
  unpeeled, halved, cored, and thinly
  sliced lengthwise

1 scallion, trimmed and thinly sliced

2 oranges, supremed, with excess juice
  separate from segments (page 121)

2 small tomatoes (preferably 1 red,
  1 yellow), chopped

2 tablespoons (30 ml) fresh lime juice

2 tablespoons (30 ml) olive oil or
  pumpkin seed oil

Pepper

3 tablespoons (12 g) pumpkin seeds,
  toasted

In a dry skillet over medium-high heat, toast the cumin seeds until they start to smoke. Turn to low heat and add the coriander, cinnamon, garlic powder, and chipotle pepper. Cook for an additional 1 minute, cool, and grind until fine with 1 teaspoon (6 g) of salt in a spice grinder.

In a pot of salted, boiling water, blanch the nopales for 1 minute or until just softened. Immediately plunge into an ice bath to stop the cooking. Dry off and combine with the chayote, apple, scallion, orange segments and juice, tomatoes, lime juice, and spice blend. Let stand for 30 minutes or until the chayote wilts.

Place salad in a colander over a bowl to collect the juices from the salad. Add the juice to a saucepan, bring to a boil, and reduce to 3 tablespoons (45 ml) (it should be syrupy). Remove from heat, whisk in oil, and season with salt and pepper. Allow dressing to cool before tossing with the salad. Sprinkle with pumpkin seeds and serve.

Yield: 4 servings

Chayote is in season in the summer/fall months, overlapping its season with apples and nopales, making this combination a go-to from July through October.

"LIFE OF THE PARTY"

AVOCADO + JICAMA + CHILE PEPPER

Avocado's creamy, luscious quality contrasts wonderfully with the slightly sweet, very crunchy jicama. Both avocado and jicama have relatively mild flavors, so chile peppers spice up this combination. It can be rather simple to put together, as avocado and jicama are both best served raw. The cooking element comes in with treatment of the chile pepper—raw for a strong kick or cooked to soften the flavor.

Let avocado lead the flavor pack in a smooth soup garnished with a jicama-chile salsa. Let jicama play more of a role in a crunchy salad. Or leave roasted poblano peppers whole on a fried fish sandwich with avocado and jicama as relish. These flavors accommodate chicken, pork, beef, and seafood very well.

AVOCADO

If left at room temperature, avocados continue to ripen after they are picked. A hard avocado just doesn't compare to a ripe one, so don't be tempted to use one that feels hard to the touch. It should yield to gentle pressure. Once ripe, store in the refrigerator.

JICAMA

Crunchy, sweet, and high in fiber, jicama is a healthy addition to many dishes. Jicama root can be large, more than you'll likely need for a particular recipe, so use thin slices as a substitute for corn chips, with other vegetables for crudités, or as a crisp addition to tuna salad. Also, use it as a substitute for water chestnuts.

CHILE PEPPER

Use the chile in this combination not just for heat but also to bring out the sweetness of the jicama and lend a beguiling note to the mild avocado. Just about any pepper can work. However, avoid habañeros unless using only a few small slivers; their heat is intense.

"LIFE OF THE PARTY" RECIPE

Not Your Mama's Guacamole

This is not a typical guacamole. It's chunkier and with more crunch because of the jicama. Serve with fish tacos, grilled shrimp, or as a dip.

1 medium **chile pepper** (such as aji amarillo or jalapeño),
    split lengthwise and seeds removed
2 cloves garlic, peeled
½ small onion
Juice of 1 lime
2 tablespoons (2 g) chopped cilantro
1 tablespoon (20 g) honey
1 medium tomato, diced
1 **ripe avocado**, diced
½ cup (65 g) **diced jicama**
Salt and pepper

Preheat oven to 350°F (180°C, or gas mark 4). Place chile pepper, garlic, and onion on a sheet pan and roast for 10 minutes or until slightly soft. Let cool slightly and chop.

In a small bowl, mix lime juice, cilantro, and honey. Stir roasted peppers, garlic, onion, tomato, avocado, and jicama together in a large bowl. Drizzle lime juice mixture over, season with salt and pepper, and stir gently to combine. Serve immediately.

Yield: 2 cups (250 g)

Because avocados oxidize and turn brown when exposed to air—even with the addition of lime juice— don't mix this together early. Assemble all of the other ingredients ahead of time and then add the avocado just before serving.

"THE ALLURE OF MEDITERRANEAN CAVIAR"

BOTTARGA + GARLIC + LEMON

This straightforward yet flavorful combination captures a distinctly Sardinian flavor. The salty, pungent flavor of bottarga (a dried and salted fish roe) is matched with the power of garlic and the tart acid of lemon. The combination comes together easily, as bottarga needs little in the way of preparation, yet it achieves a sophisticated flavor.

The application recipe prepares a simple spaghetti that packs a distinctively tasty punch. Also look to the combination to add flavor to risotto, shellfish, and vegetables such as mushrooms and eggplant.

BOTTARGA

Use this condiment with restraint because of its strong flavor. It is usually sold as a chunk and then shaved or grated to add to dishes. Although it can be expensive (hence the nickname "Mediterranean caviar"), a little goes a long way, and it keeps for months wrapped in the refrigerator. Choose tuna bottarga for a stronger, more pungent taste or gray mullet for a slightly milder flavor.

GARLIC

Because this combination pairs garlic with another pungent taste, we recommend cooking it. Sautéing it will soften its impact, allowing the bottarga to shine yet supporting the overall flavor. Use roasted garlic for its creamy, mild texture and mix it with grated bottarga and lemon zest as a spread for canapés or grilled vegetables.

LEMON

This combination needs lemon to tweak the flavors and make them come alive in your mouth. Add only a squeeze of its juice to contribute the acid element vinegar offers. This alone accomplishes much toward a more dynamic-tasting dish. Add finely grated lemon zest for flavor that will be much appreciated.

"THE ALLURE OF MEDITERRANEAN CAVIAR" RECIPE

Extra-Special Spaghetti

Although this is a breeze to prepare, you'll marvel at the richly flavorful result.

1 package (1 pound, or 455 g) spaghetti

6 tablespoons (90 ml) olive oil

5 cloves **garlic**, minced

1 teaspoon (1 g) red pepper flakes

Juice of ½ lemon

2 tablespoons (28 g) unsalted butter

1 bunch flat-leaf parsley leaves, chopped

2 ounces (50 g) **bottarga**, shaved

Zest of 1 lemon

Bring a large pot of salted water to a boil to cook the spaghetti.

While the pasta cooks, heat the olive oil in a skillet over medium heat. Add the garlic and pepper flakes and cook just until the garlic softens and becomes fragrant (do not let it brown). Remove from heat.

Drain spaghetti when it is al dente, after 8 to 12 minutes, and return it to the pot. Scrape in the oil mixture with a rubber spatula. Add the lemon juice, butter, and parsley. Toss to combine and divide into serving bowls. Top each bowl with shaved bottarga and lemon zest and serve immediately.

Yield: 4 servings

For variation, add shrimp or mussels to this dish. Also, use any type of pasta you wish.

"BREAKING OUT OF THE SHELL"

COFFEE + CARDAMOM + PISTACHIO

Traditional Middle Eastern cuisine frequently pairs the full, rich bitterness of coffee and the distinctive spiciness of cardamom. Add another Middle Eastern component—the pistachio—and you have an exotic flavor trio.

The application recipe prepares a creamy cardamom-coffee pots de crème with a pistachio brittle. Other options for the combination: a cardamom-spiced cup of coffee with pistachio biscotti, an infused custard turned into ice cream, or a pistachio-cardamom coffee cake with espresso crème fraîche. These flavors go great with orange, honey, cream, and saffron.

COFFEE

To infuse coffee into cream, as the application recipe does, use deeply roasted beans. To add coffee flavor as a prepared liquid for cooking, go for espresso for its richly concentrated flavor. French press or strongly brewed drip coffee work as well. With strong flavors at work here, you'll need strong coffee to keep its flavor relevant in the combination. In a pinch you can use instant espresso powder, but do not use regular instant coffee—it just isn't flavorful enough.

CARDAMOM

When using whole pods in liquids, as the application recipe does, be sure to break them open to release the inner seeds and flavor. Using ground cardamom imparts a significant amount of flavor, so start with a small amount and add more to taste. Try to use pods when possible. For ground, toast your own pods and grind in a spice grinder.

PISTACHIO

Pistachios in their shells are great for snacking, but shelling a large quantity can be time consuming, so keep a stash of shelled pistachios in your freezer for freshness. Toast them to accentuate their flavor but go lightly to preserve their naturally green color. For a substitute, use almonds (though it won't quite be the same).

"BREAKING OUT OF THE SHELL" RECIPE

Coffee-Cardamom Pots de Crème with Pistachio Brittle

This is not only a wonderful contrast of flavors, but also of textures—the creamy spoonful of chilled custard with the crunchy bite of nutty brittle. You can prepare both components the day before serving.

Pistachio Brittle

2 tablespoons (28 g) unsalted butter, plus more for spatula, divided

1⅓ cups (270 g) sugar

¼ cup (85 g) light corn syrup

½ cup (120 ml) water

¼ teaspoon baking soda

¼ teaspoon salt

Finely grated zest from 1 orange

1½ cups (185 g) **shelled pistachios,** toasted

Pots de Crème

¾ cup (45 g) **dark roasted coffee beans**

2 tablespoons (12 g) **cardamom pods**

1½ cups (355 ml) heavy cream

¾ cup (175 ml) milk

6 egg yolks

⅔ cup (135 g) sugar

🍴 This is a great dessert for the winter months, when fresh produce is more scarce and the oven's heat is welcome. Store any extra pistachio brittle in a tightly closed container for up to 2 weeks.

To prepare the pistachio brittle: Line a sheet pan with a nonstick liner (or butter the bottom and sides). Butter the back of a spatula and set both within reach.

In a heavy saucepan, combine the sugar, corn syrup, water, and butter. Cook over medium heat, stirring to combine, and then leave it untouched as it boils. Lift and swirl the pan to evenly cook, if necessary, but do not stir. Let it cook until it reaches a light brown color and remove from the heat.

Working quickly, stir in the baking soda, salt, orange zest, and nuts with a rubber spatula. Pour onto prepared sheet pan and use the buttered spatula to spread as thin as possible. Do not touch with your bare hands; it is molten hot! Let brittle cool completely and break into small pieces.

To prepare the pots de crème: Preheat the oven to 300°F (150°C, or gas mark 2). Roughly chop the coffee beans and cardamom pods by hand or in food processor (do not grind). Place in a saucepan and add the cream and milk. Bring to a simmer, remove from heat, and cover with plastic wrap. Let the cream infuse for 30 minutes.

Strain through a fine mesh strainer, discard solids, and add cream mixture back to a clean saucepan. Bring to a simmer over medium heat. In a medium-size bowl, whisk together yolks and sugar for 1 minute until lightened. Add a small ladle of the hot cream to the eggs and whisk. Add another couple ladles of cream and whisk.

Add the egg mixture back to the simmering cream in a steady stream, continually whisking the entire time. Cook, stirring and scraping the edges with a rubber spatula, until thickened (the custard should coat the back of a spoon).

Strain through a fine mesh strainer and divide among four 4- to 6-ounce (120 to 175 ml) ramekins. Place in a roasting pan and fill two-thirds full with hot water. Bake for 40 minutes or until the custards are just set but still slightly jiggle in the center. Remove and allow to cool for 20 minutes in the roasting pan and then remove from the pan and chill in the refrigerator.

When they are cold, cover with plastic wrap until serving. Serve with a jagged piece of brittle stuck in each pot de crème.

Yield: 4 servings

"A SWEET BIT OF ITALIA"

OLIVE OIL + CITRUS + WINE

Olive oil is a staple in many recipes, but unless you are dipping bread in it, it rarely gets top billing. Typically, its heart-healthy, smooth flavor cooks other ingredients or as a component in a vinaigrette. Here it has a new role: star of the baked show. When it replaces butter in cakes, it lends a subtle but delicious flavor with zingy citrus and fruity wine, creating a mouthwatering tenderness.

The application recipe prepares an orange-Sauternes pound cake. This combination goes great with peaches, figs, honey, and nuts.

OLIVE OIL

Personal preference should guide your choice in an olive oil for baking. Extra-virgin olive oil is highly esteemed (and even has its own acronym), but the pure flavor extracted from the first pressing of the olives gives a fruitier flavor. This is great, given the use of citrus in the combination. However, regular olive oil has an overall lighter flavor that makes an equally good cake—and it is less expensive.

CITRUS

The classic citrus flavors used here are lemon and orange. The zest in particular adds an intense citrus flavor to baked goods. Be sure to use a micrograter to extract the zest without any of the bitter white pith. Use the juice as well and even a citrus liqueur (such as Cointreau) to boost the citrus. For a different twist, try grapefruit.

WINE

Consider one of two different choices for this combination: Sauternes or sherry. Sauternes is a French dessert wine made from Sémillon, Sauvignon Blanc, and Muscadelle grapes affected by botrytis (the "noble rot"), resulting in a concentrated, uniquely flavored wine. Sherry comes primarily from Palomino grapes and is fortified with brandy. Choose a medium sherry, such as amontillado, for this combination.

Olive Oil Pound Cake with Citrus and Sauternes

Garnish this tender, flavorful cake with raspberries. Wrap leftovers with plastic wrap-—the flavors actually get better with time.

Butter and flour to grease pan

3 cups (375 g) all-purpose flour

2 teaspoons (9 g) baking powder

¼ teaspoon salt

2 cups (400 g) sugar

1 cup (235 ml) **olive oil** (extra-virgin or regular; see above discussion)

Zest from 1 **orange**

Zest from 1 **lemon**

5 eggs

1 teaspoon (5 ml) **orange-flavored liqueur** (such as Cointreau), or **orange flower water**

1 cup (235 ml) **Sauternes wine**, divided

Bake in two small loaf pans as an alternative, so you can freeze one for a future treat.

Preheat the oven to 350°F (180°C, or gas mark 4) and butter and flour a tube or bundt cake pan.

In a bowl, whisk together flour, baking powder, and salt. In the bowl of a standing mixer with the whisk attachment, mix the sugar, olive oil, and zests on high speed for 1 minute. Add the eggs, one at a time, waiting until the prior egg fully incorporates before adding the next. Beat on high for 5 minutes until mixture is thick and white.

Add one-third of the flour mixture and mix on low speed until just combined. Add the liqueur and half the wine, mixing until just combined. Repeat the process, ending with the last one-third of the flour. Take care not to overmix the batter as it will compromise the cake's texture.

Pour batter into the prepared pan and bake for 1 hour or until a toothpick emerges clean from the center of the cake. Allow to cool for 20 minutes in the pan and then carefully unmold. Allow to cool completely on a wire rack and keep tightly wrapped in plastic wrap for up to 3 days.

Yield: 1 cake, 8 to 10 servings

"A TRANSFORMATIONAL EXPERIENCE"

OKRA + FENNEL + CORIANDER

Okra is the kind of vegetable people want to like, but unless they have fond childhood memories of eating it, they usually can't get past the texture—as in, "they are so slimy."

Okra really can be slimy when cooked slowly, but—and this is when a cooking application can really transform an ingredient—in this case, it gets fried to a nice crispy texture. Ground fennel and coriander seeds give an Indian-spiced flavor to the combination. Turn this into a stew by slow cooking and letting the slime factor work for you as a natural thickener in the dish. Serve with Indian curries, lentils, rice, or beans.

OKRA

A favorite in the South, this vegetable commonly appears in gumbo and is used both for flavor and as a natural thickener. It can be pan- or deep-fried for crispiness. Cooking it slowly with other liquids gives it a slippery feel but thickens the surrounding liquid. The smaller the okra, the more flavorful—avoid any longer than four inches (10 cm), if possible.

FENNEL

Fennel's most intense flavor comes when it is raw and crunchy. Sauté and simmer it to soften or roast it until it practically melts. Candy it for sweetness. Its seeds are ground and used as a flavoring in this application recipe. However, you can also dredge fennel bulb and fry it just like the okra—just be sure to blanch it for a few minutes and dry it off before proceeding.

CORIANDER

The coriander seed has a nutty flavor accentuated by roasting. Buying whole seeds, roasting them, and grinding them gets the maximum flavor possible. Ground coriander is acceptable; just be sure to use it soon after buying as its flavor diminishes over time.

"A TRANSFORMATIONAL EXPERIENCE" RECIPE

Crispy Spiced Okra

Serve this as a casual appetizer or make a meal out of it by adding rice and a salad. If you are inclined toward a condiment for dipping, try Mango Ketchup (page 173) or Pumpkin Seed Salsa (page 215).

Oil for frying (such as canola or corn)
¼ cup (20 g) **coriander seeds**
¼ cup (23 g) **fennel seeds**
½ cup (63 g) all-purpose flour
½ cup (70 g) cornmeal
1 teaspoon (6 g) salt
⅛ teaspoon cayenne pepper
2 cups (460 g) plain yogurt
2 tablespoons (30 ml) milk
1 pound (455 g) **okra**, stems trimmed
   off, sliced in half lengthwise

Place oil in a large, heavy-bottomed pot to a depth of 2 inches (5 cm). Over medium-high heat, bring to 350°F (180°C).

Put coriander and fennel seeds in a spice grinder (or cleaned-out coffee-bean grinder) and grind to a fine powder. Mix with flour, cornmeal, salt, and cayenne pepper. Transfer to a shallow container or pie plate.

Whisk together yogurt and milk in a medium bowl. Drop a handful of sliced okra into the yogurt mixture, remove the pieces one at a time, and coat in the seasoned cornmeal. Drop them individually into the hot oil, being careful not to overcrowd the pan. When they turn golden and crisp, remove with a slotted spoon and drain on paper towels. Repeat until all okra is fried. Serve immediately.

Yield: 4 servings

A tip about serving okra: Don't ask, "Who feels like okra tonight?" It can be a hard sell. Instead, prepare it properly to both please and surprise your friends and family.

"FULL-FRONTAL FLAVOR"

SQUASH + RED CURRY + NUT

Winter squash offers a hearty flavor and creamy texture that embraces the dynamic flavor profile of red curry paste. The crunch and depth of nuts offer a pleasant textural bite.

The application recipe prepares a creamy curry, perfect over steamed rice. Other options for the combination: Season acorn squash with curry, roast in the oven, and serve with cashews and lentils; or purée cooked pumpkin for a creamy curry soup garnished with almond slivers. These flavors are well suited for chicken, tofu, shrimp, and legumes.

SQUASH

Butternut squash has a great hue and wonderful flavor and requires the least amount of work to manage. If you plan to purée the cooked squash, roast it whole to avoid the cumbersome task of cutting up a raw squash.

RED CURRY

A red curry paste is an easy way to keep this seasoning in your refrigerator for an extended period of time, and it tends to have more flavor than powders. The bite of red curry will stand up to any cooking preparations for the combination, so remember to go easy in the beginning.

NUT

Toasting and roasting nuts intensifies their flavors—a needed effect in this application so they can hold their own with the luscious brown butter. They primarily provide a textural note in this dish. With the strong curry flavors, use almost any type of nut. Those traditionally used with curry include peanuts and cashews. Also look to pine nuts and macadamias for a different flair.

"FULL-FRONTAL FLAVOR" RECIPE

A Fall Curry to Remember

While this recipe calls for butternut squash, you will get nice results with acorn squash or sugar pumpkin as well.

1 teaspoon (2 g) whole cumin seeds

1 teaspoon (2 g) whole coriander seeds

1 cinnamon stick

½ teaspoon ground nutmeg

½ teaspoon whole peppercorns

1 tablespoon (15 ml) vegetable oil
  (such as canola)

1 onion, diced

2 cloves garlic, finely chopped

1 inch (2.5 cm) fresh ginger, peeled and
  finely chopped

2 dried chiles de arbol, crushed into
  small pieces

4 cups (560 g) **butternut squash**,
  peeled, seeded, and cut into ¾-inch
  (2 cm) cubes

1 can (14 fluid ounces, or 425 ml)
  coconut milk

1 tablespoon (15 ml) rice vinegar

1 tablespoon (15 ml) plum wine
  (optional)

2 tablespoons (30 ml) soy sauce

1 tablespoon (15 g) **red curry paste**

Salt and pepper

Sriracha sauce (optional, for
  additional heat)

¼ cup (4 g) chopped fresh cilantro leaves

¼ cup (25 g) chopped scallions

½ cup (73 g) chopped **roasted peanuts**

In a large skillet over medium heat, cook cumin, coriander, cinnamon, nutmeg, and peppercorns, stirring frequently, until lightly toasted and fragrant. Add to spice grinder (or cleaned-out coffee-bean grinder) and pulse to a fine powder.

In the same large skillet, heat the oil and add the onion, garlic, ginger, and chiles. Cook until soft and translucent. Add the spice blend and squash; stir to incorporate. Cook over medium heat until the squash begins to soften.

To the pan, add coconut milk, vinegar, wine (if using), soy sauce, and curry paste. Cook until the curry dissolves and the squash softens but still holds its shape. Taste and adjust seasoning with salt, pepper, and Sriracha, if necessary. Sprinkle the top with cilantro, scallions, and peanuts. Serve immediately, over rice if desired.

Yield: 4 servings

Use seasonal squashes with curry to warm up on cooler evenings. This recipe could easily accommodate tofu, chicken, or shrimp.

"INDIAN TART AND SWEET"

CURRY + APPLE + COCONUT

Piquantly spiced curry meets tart apple and sweet coconut for a combination of interesting depth. These exotic and familiar flavors come together in a variety of ways.

The application recipe prepares a version of the richly spiced Mulligatawny stew. Other options include a chicken curry with apple-coconut chutney or a spicy shrimp curry with coconut milk and shaved apple. These flavors taste great with chicken, pork, shellfish, seafood, tofu, raisins, or yogurt.

CURRY

In India, the flavors that compose curry can be quite diverse, but in Western cuisine, it has become a generic description of a spice blend. The spices generally include turmeric (giving it the characteristic yellow color), cumin, coriander, and fenugreek (used for its tang and intensity), among others. The powder form (also called garam masala) is most easily obtained; purchase it in small quantities, as its flavor quickly fades.

APPLE

Use raw apples to get the greatest texture and crisp flavor. For raw applications in this combination, seek out Pink Lady or Gala apples. Cooking softens the intensity of the apple's flavor and texture. Tart Granny Smith apples balance the sugars that increase as apples cook. To let apple dominate, make an apple-coconut chutney to serve with curried chicken or shrimp.

COCONUT

The easiest way to use coconut in this combination is as coconut milk—a liquid in which to cook rice, a base for curry sauces, or in custards. Flaked coconut also can add a nice textural interest to these dishes or in baked applications. Toast it to deepen its flavor; use unsweetened flaked coconut for the purest flavor.

"INDIAN TART AND SWEET" RECIPE

Mulligatawny Soup

Make a meal out of this by adding a loaf of bread and a hoppy India Pale Ale beer.

1 tablespoon (15 ml) olive oil

1 onion, diced

4 cloves garlic, peeled and diced

2 **Granny Smith apples**, peeled, cored, and diced

Cooked chicken meat (8 to 12 ounces [225 to 340 g]), chopped

½ cup (120 ml) dry white wine (such as Sauvignon Blanc)

1 quart (945 ml) chicken broth

2 bay leaves

1 tablespoon (6 g) **curry powder**

1 cup (165 g) cooked white rice

1 cup (235 ml) unsweetened **coconut milk**

Salt and pepper

In a large pot, heat the olive oil over medium heat and lightly sauté the onion, garlic, and apples. Add the chicken, wine, broth, bay leaves, and curry powder.

Stir together, bring to a low simmer, and cook for 5 minutes. Add the rice and coconut milk and season with salt and pepper. Bring to a simmer and serve hot. Remove bay leaves before serving.

Yield: 4 servings

This dish benefits from being made the day before, so its flavors have a chance to develop.

CHAPTER 7

Decidedly Decadent

There is a time and a place for indulgence, for abandoning a sense of moderation, calorie counting, and carb reduction. A time when we let go of what we are supposed to eat and go for the biggest, richest mouthful of flavor possible. Sometimes, we need to mark a special event such as a birthday or anniversary, and sometimes we need to go over the top just to remind ourselves that we are, indeed, living, not merely products of our routine.

If one quality unites the recipes in this chapter, it is richness. You reach for these kinds of recipes when you want to pull out all the stops. They are not intended for everyday dining (but really, that's up to you). We designed these dishes to provoke the response, "Wow! Now that's eating." And in the spirit of engaging in most things decadent, something seems "naughty" about eating them.

Some of the recipes are savory, others are sweet. The savory recipes often look to luxury ingredients to provide rich flavor or an indulgent aspect. Speckled Butter (page 237) combines truffles with butter (the higher the fat, the better) and black pepper for a creamy, aromatic spread. Several recipes feature foie gras, perhaps one of the richest, silkiest mouthfuls one can experience. Look to Foie Gras to the Tenth Degree (page 243), where we combine this ingredient with bacon, for a true representation of the meaning behind this chapter. Other ingredients featured for their over-the-top savory experience include caviar and bone marrow.

We would be overlooking the obvious if we failed to address the sweet aspect of culinary decadence. Chocolate naturally figures into some of these recipes, especially one of the most popular desserts at our restaurant—Chocolate Soufflé with Kahlua-Chocolate Sauce and Whipped Cream (page 256). We also combine chocolate with goat cheese in Heavenly Chocolate Fondue with Strawberries (page 271) for a unique, smooth dessert. Or for a more casual but equally rich treat, try the Triple Threat Milkshake (page 275), which brings into the mix peanut butter and malt.

But chocolate is not the only rich sweet—Liquor-Laced Old-School Eggnog (page 273) will knock you over with its intensity. And if you are looking for a decadent sweet dessert that is also fancy, try Lemon-Raspberry Meringue Nests (page 268), a version of a Pavlova dessert.

Whatever you do, be sure to try something from this chapter. After all, life is short, and sometimes we need to slow down enough to savor it. Dip your toe in the decadent waters and enjoy every bite.

"AROMATIC VELVET"

TRUFFLE + PEPPER + BUTTER

The richly aromatic truffle gives this combination its distinctive earthy edge. Add the spice of freshly ground pepper and you have quite a flavor. The missing factor is an ingredient that rounds out these unique flavors and adds a velvety richness to serve with food. Enter butter! Because these are all powerful flavors, let butter be the primary vehicle to deliver the truffle and pepper. These flavors go great with potatoes, pasta, and risotto.

TRUFFLE

Because of truffles' pungent flavor—white truffles have a stronger flavor than black— and high cost, cooks use them sparingly in food. They are available seasonally fresh or jarred in a light brine. Truffle-flavored oils and butters also are readily available.

PEPPER

The fiery sibling of salt, pepper (from peppercorns) is most flavorful when freshly ground. Black has the most kick of the various colored peppers, but white, green, and pink also are flavorful and visually interesting.

BUTTER

Butter makes food taste better, so don't be afraid to use it in your cooking. It's available with or without added salt, but as a rule, stick with unsalted so you can control your own seasoning. For compound butters such as that in the following recipe, consider using a high-fat, European-style butter such as Plugrá.

"AROMATIC VELVET" RECIPE
Speckled Butter

This homemade truffle butter is like butter gold. Use it to tweak sauces, finish risotto, toss with hot French fries, or even drizzle over popcorn.

1 pound (455 g) **salted butter**, softened

2 tablespoons (28 g) **black truffle** peelings

2 tablespoons (30 ml) **truffle oil** (if possible, use 1 tablespoon [15 ml] black and 1 tablespoon [15 ml] white)

1½ teaspoons (3 g) **black pepper**, freshly ground

In the bowl of a standing mixer, beat together all ingredients with the paddle attachment. Use immediately or package for longer storage.

Yield: 2 cups

This freezes well, so consider separating your batch into small containers to load into the freezer. Then you can pull them out individually to add a special touch to any dish.

"LUSCIOUS, MEET SWEET AND SAVORY"

LIVER + PEA + SAGE

Liver has a strong, unique flavor; some think they don't like it, but that impression usually comes from eating some that was poorly prepared. At its best, it is silky, rich, and wonderfully complemented by the sweet, vegetal pea and herb-forward sage.

For a simple preparation of the combination, sauté the liver with fresh peas and a sprinkle of chopped sage. The application recipe takes the refined route, preparing layers of the ingredients in a rich, flavorful savory parfait. This is the perfect recipe to prepare for those who think they don't like liver.

LIVER

Chicken livers are by far the most easily accessible and least expensive liver to purchase. They make a nice choice for this combination. Try to find duck livers if you can; they are harder to get, but they make for a richer liver flavor. Foie gras is specialty duck or goose liver that results from force-feeding the bird. It is much larger and richer than the average liver. It can be used in the combination, but sear it to highlight its wonderful, luscious qualities.

PEA

The world of peas includes the shelled variety (with inedible pods) and pod peas, which you can eat the whole (sugar snap peas are an example). For this combination use shelled peas. Don't even get close to canned peas, but keep a stash of frozen around to easily add to your dishes. Peas purée nicely for use in soups and smooth sauces.

SAGE

This highly aromatic herb is easy to grow in home gardens and will usually winter over, even in colder climates. To extend the use of fresh sage, cover in olive oil and keep refrigerated for up to two months. Fresh sage flash-fries extremely well and makes a crisp, flavorful garnish.

"LUSCIOUS, MEET SWEET AND SAVORY" RECIPE

Chicken Liver Mousse with Pea Purée and Sage Cream

This complex, layered first course requires several steps but is worth it with its lovely presentation and rich flavors. Be sure to make the layers in the order in which they are given.

Pea Layer

3 tablespoons (42 g) unsalted butter, divided

½ small onion, finely chopped

1½ cups (225 g) **peas**, fresh (frozen and thawed will work)

¼ cup (60 ml) vegetable broth

½ bunch fresh parsley, leaves only, blanched in boiling salted water and shocked in ice bath

Salt and white pepper

Chicken Liver Mousse Layer

¼ cup (55 g) unsalted butter, divided

1 pound (455 g) **chicken livers**

1½ teaspoons (1 g) fresh thyme leaves (lemon thyme preferred)

2 cloves garlic, finely chopped

1 small onion, finely chopped

1 bay leaf

2 tablespoons (30 ml) brandy

Salt and pepper

To prepare pea layer: In a skillet, heat 1 tablespoon (14 g) butter over medium heat; add onion. Cook until onion is translucent. Add peas and cook until peas soften. Transfer to a blender and add vegetable broth, parsley, remaining butter, salt, and white pepper. Blend until smooth.

Divide mixture among four 6- to 8-ounce (175 to 235 ml) ramekins or wide-mouth glasses, smooth the top, and refrigerate while preparing the chicken liver mousse.

To prepare liver mousse: In a large skillet over medium heat, cook 1 tablespoon (14 g) of butter with liver, thyme, garlic, onion, and bay leaf. Cook until the livers brown lightly and the onion and garlic soften.

Deglaze the pan with the brandy and cook off for 2 minutes. Remove the bay leaf and place all other pan ingredients into a blender; allow to cool in the blender for 5 minutes. Season with salt and pepper and add remaining butter. Blend until creamy. Add to the ramekins, on top of the pea layer, and smooth the top. Chill in the refrigerator while preparing sage cream.

(continued on next page)

Sage Cream

¼ sheet gelatin (or ½ teaspoon
powdered)

2 tablespoons (30 ml) cold white wine
(something sweet and spicy such as
Gewürztraminer)

2 cups (475 ml) heavy cream

14 leaves **fresh sage**, divided

1 bay leaf

1 tablespoon (20 g) honey

1 cup (235 ml) peanut oil
(or canola oil)

**This is a refined, complex
dish that makes a rich first
course. You can make this up
to 1 day before serving.**

To prepare sage cream: In a small bowl, soften the gelatin in the wine and set aside. In a saucepan, combine cream, 6 sage leaves, bay leaf, and honey. Bring to a simmer and reduce to ½ cup (120 ml) cream. Strain cream into the softened gelatin and stir to combine. Pour on top of mousse layer and chill to set.

Bring peanut oil to 300°F (150°C) in a small saucepan. Add remaining 8 sage leaves and fry until crisp. Remove with slotted spoon, drain on paper towels, and sprinkle with salt.

Before serving, place 2 fried sage leaves on top of each parfait. Serve chilled.

Yield: 4 parfaits

"THE DECADENT DEEP END"

BACON + APPLE + COGNAC

This intermingling of rich, contrasting flavors can be quite decadent when used with rich meats in particular. The tart, sweet apple lightens and elevates the salty, fatty bacon. Cognac adds a deep, rich flavor base that further intensifies the combination.

The application recipe goes off the richness charts by adding foie gras. Other opportunities include using the ingredients to flavor a butternut squash soup or as a stuffing for poultry.

BACON

This combination features dynamic flavors, so feel free to experiment with different types of bacon—such as peppered or brown-sugar baked. Also look into pork belly, the same meat used in sliced bacon but left in large cuts. It is a favorite for restaurant chefs for its amazing texture, juiciness, and rich flavor. Cognac-drenched apples make a nice addition to a wedge of braised pork belly.

APPLE

In general, look for a tart-flavored apple to cut through the rich, strong flavors. Varieties such as Granny Smith and Braeburn work well. Cooking the apples with the Cognac and bacon allows for great flavor integration, but to accent the apple flavor more and add texture and flavor, use it raw. Also insert apple flavor into a dish by using apple juice or cider as a reduction or cooking liquid.

COGNAC

Cognac enjoys the status of mother of all brandy. It is produced in the Cognac area of France, distilled from grapes, and aged to a complex, rich, bold flavor. High-quality brandies and Armagnac also work for this combination. Because most recipes call for relatively small amounts of this, invest in a nice bottle and save the rest to sip after dinner.

"THE DECADENT DEEP END" RECIPE

Foie Gras to the Tenth Degree

If you are not accustomed to searing foie gras, watch it very carefully. Your pan needs to be very hot before you start cooking, but the liver's fat will render quickly as it cooks so stand guard.

2 slices **bacon**, cut into ½-inch (1¼ cm) pieces
1 shallot, finely chopped
1 **Granny Smith apple**, peeled, cored, and diced
½ cup (55 g) dried bread crumbs
¼ cup (60 ml) **Cognac**
¼ cup (60 ml) sweet white wine (such as Muscat)
1 dried bay leaf
¼ cup (60 ml) heavy cream
4 slices foie gras, each ¾-inch (2 cm) thick
Salt and pepper

In a large skillet, cook bacon until not-quite crisp. Add the shallot, apple, and bread crumbs. Cook until the crumbs start to brown, the bacon is crispy, and apples soften.

Deglaze the pan with Cognac and wine and add the bay leaf and cream. Stir to combine and cook until you reach desired sauce consistency. Keep warm while cooking foie gras. Remove bag leaf.

With a sharp knife, make shallow crosshatches over the surface of each piece of foie gras. Season both sides with salt and pepper. Over high heat in a hot skillet, sear foie gras crosshatch-side down first (so that when serving it will be up). Cook for 1 to 2 minutes per side and then drain on paper towels. The outside should be dark brown, but the inside should be only warm and not cooked through—squeeze the sides to check that they yield and are not firm. Immediately serve foie gras topped with sauce.

Yield: 4 servings

🔲 These are fall-season flavors, fitting to eat when the hot summer months are far behind and a chill has settled into the air.

"FIERY SWEET"

QUINCE + MANGO + CHILE PEPPER

The exotic flavor and aroma of quince comes out of its fall slumber, awakens with the tropical sweet mango, and receives a nice dose of fire with chile pepper. These flavors stimulate the palate with their tart-sweet-spicy complexity.

The application recipe brings the combination to rich squab meat—preparing both a Thai chutney and pan sauce. Because quinces respond so well to cooking, consider making a jam out of the combination or a dessert of a quince tarte Tatin with a mango-chile crème anglaise. These flavors taste great with poultry such as duck, quail, or chicken, as well as with vegetarian rice dishes.

QUINCE

These hard fruits look like a cross between an apple and a pear and have a penetrating perfume and a sweet-tart flavor. You must cook them before enjoying them—poach, roast, or bake them to soften and bring out their sugars. They are loaded with pectin (a natural jelling agent) and as a result, make a great jam. Quince paste (also called membrillo) is readily available in jars and is a nice addition to a cheese plate.

MANGO

Fresh, ripe mangoes provide the best flavor and texture for the tart quince. If the mangoes don't look great, consider using mango juice (Odwalla makes a nice one) to cook the quince in (roast it in the oven or slow simmer it in a skillet). Or try slow-cooking chopped dried mango with the quince to soften and blend the flavors.

CHILE PEPPER

A Thai chile pepper is a nice choice for its medium heat and bright red color contrast. The birds-eye pepper is the most common variety, and though it is small, it packs strong heat. Cooking the chile softens the blow of the heat; reduce the quantity if using birds-eye chile peppers, because they contain a considerably stronger heat. A chile de arbol makes a nice alternative and can be quite hot if fresh; it is also sold dried.

"FIERY SWEET" RECIPE

Succulent Squab with Quince-Mango Chutney

Squab breasts should be served medium rare for the best flavor and texture, so if using the whole bird, cook the leg and thigh separately (breast should be 130°F [54°C] internal temperature, leg/thigh should be 140°F [60°C]). You may substitute duck breasts for the squab.

Quince-Mango Chutney

1 teaspoon (2 g) chopped fresh ginger

½ ripe **mango**, peeled and seeded

Juice of 1 lemon

Salt

¼ cup (60 ml) simple syrup (page 183)

¼ cup (24 g) fresh mint leaves

1 tablespoon (15 ml) canola oil

1 small red onion, diced

4 **Thai chiles**, finely chopped

3 medium **quinces**, peeled and diced

Juice of 1 lime

Pepper

Squab

8 boned squab half-breasts, with legs/
   thighs cut separately

Salt and pepper

2 tablespoons (30 ml) canola oil

¼ cup (60 ml) semisweet wine
   (such as Riesling)

½ cup (120 ml) chicken stock

1 tablespoon (14 g) unsalted butter

You will have extra quince-mango chutney after you make this recipe. Use it on sandwiches, on bagels with cream cheese, or can it and give as a gift.

To prepare chutney: In a blender, combine ginger, mango, lemon juice, a pinch of salt, simple syrup, and mint. Blend until very smooth; set aside.

In a skillet, heat the canola oil over medium heat, add onion and chiles, and cook for 6 minutes until softened and browned. Add quinces, lime juice, and mango mixture from the blender. Cook over low heat until the quinces soften, up to 20 minutes. Season with salt and pepper, cool, and refrigerate until ready to use, up to 1 week.

To prepare squab: Preheat oven to 350°F (180°C, or gas mark 4). Salt and pepper both sides of the squab pieces. Over medium-high heat, sear in canola oil until brown, turn over, and brown other side. Remove to a sheet pan and cook in the oven until the breasts reach 130°F (54°C) and the leg and thigh reach 140°F (60°C) when tested with a meat thermometer. Allow to rest on a plate while preparing sauce.

Meanwhile over medium heat, deglaze the pan used to sear the squab with the white wine and then add the chicken stock. Simmer and cook for 10 minutes until the liquid reduces by half. Add ¼ cup (65 g) of the chutney, stirring until dissolved. Finish the sauce by adding the butter and seasoning with salt and pepper.

Serve squab topped with sauce and a dab of chutney on the side.

Yield: 4 servings

"UPTOWN SWAGGER"

EGG + CAVIAR + CHERVIL

Pairing eggs and caviar is a classic combo, and bringing in the delicate chervil adds a wonderful subtlety that elevates the entire flavor combination.

The application recipe prepares a fluffy, rich chervil-egg custard as the perfect base for the salty caviar. The many variations of egg preparation offer other options for the combination. Serve an old-school caviar with chopped hard-boiled eggs and chervil blini, crème fraîche, and red onion. Top deviled eggs with chervil and caviar or use with scrambled eggs and omelets for a special brunch.

EGG
The egg's versatility drives the options for this combination. Soft poach, hard boil, scramble, bake, or cook an egg into custards. The danger of cooking with this food (cooking it well often separates chefs from amateurs) is in overcooking it. Eggs can quickly become rubbery or cause a custard to "break," so keep a close eye on them.

CAVAIAR
For an upscale option, try a Russian caviar such as sevruga, which has a smooth and buttery flavor. Hackleback is an American caviar often compared to sevruga, with its similar flavor but lower cost. Paddlefish is another low-cost American caviar; it offers mild flavor and is considered a good "first timer's caviar."

CHERVIL
Delicate in flavor and fragrance, its flavor falls somewhere between anise and parsley. It offers an elegant addition to your cooking. Use toward the end of any cooking process to preserve the most flavor and always use the fresh leaves. To preserve the flavor (as drying renders it almost tasteless), infuse it in white wine vinegar.

"UPTOWN SWAGGER" RECIPE

Chilled Chervil Sabayon with Sevruga Caviar

This is a great first course for company, especially because you may prepare the custard up to 8 hours ahead of serving. Use demitasse coffee cups, small ramekins, or egg cups for the custard.

½ cup (120 ml) heavy cream

6 **egg yolks**

2 tablespoons (30 ml) anise liqueur (such as Pernod)

½ teaspoon salt

¼ teaspoon fresh ground pepper

1 tablespoon (1 g) **fresh chervil**

1 teaspoon (1 g) thinly sliced chives

2 ounces (57 g) **caviar** (such sevruga or hackleback)

Whip cream to soft peaks; cover and refrigerate.

Fill a saucepan one-third full with water and bring to a simmer. Find a stainless steel bowl that fits on top of the saucepan, without touching the water. In that bowl, whisk together yolks, liqueur, salt, and pepper until they are light and foamy. Set bowl over saucepan and continue to whisk constantly for 2 minutes until the eggs thicken. Take care to keep the whisk moving on all areas of the bowl so that the edges do not curdle.

Remove from heat and continue to whisk for 1 minute to bring down the temperature. Fold in the whipped cream, chervil, and chives. Divide among four small serving dishes. Refrigerate for a minimum of 2 hours or up to 8 hours. Top with caviar and serve.

Yield: 4 servings

For an appetizer for a casual gathering, serve these in small shot glasses with tiny spoons.

"BURST OF SENSUALITY"

CHOCOLATE + LIME + CREAM

Chocolate and citrus are a familiar combination, but usually it's orange that's highlighted. Lime as the accent flavor provides an unexpected burst, and adding cream brings richness as well as versatility.

The application recipe is a twist on hot chocolate. Other possibilities include chocolate lime pots de crème, chocolate tart with lime cream, chocolate cupcakes with lime buttercream, or white or dark chocolate–lime mousse.

CHOCOLATE

Many recipes for hot chocolate call for both cocoa powder and chopped solid chocolate to increase the flavor intensity. Milk chocolate, which has more sugar, makes a sweeter drink. When trying to understand cocoa percentages listed on chocolate, remember that the higher the number, the less sweet the chocolate. For a good multipurpose semisweet chocolate, pick something between 50 and 60 percent cocoa.

LIME

Be assertive when adding lime to chocolate. Often both the zest and the juice will be needed to break through chocolate's richness. Key limes, smaller and juicier with a higher acidity and distinct aroma than typical (Persian) limes, are a nice choice if you can find them. Don't substitute the bottled variety, however—it tastes different and doesn't yield the zest.

CREAM

Cream is often the base you will infuse the lime flavor in. Choose a heavy cream for maximum richness or lighten it up with milk. This recipe uses sweetened condensed milk, which starts as cow's milk but has water removed and sugar added. It makes a particularly luscious drink.

"BURST OF SENSUALITY" RECIPE

Key Lime Hot Chocolate

Make this delicious recipe in advance and keep refrigerated until serving. Be sure to warm it up slowly and blend it again prior to serving.

1 cup (235 ml) water

Zest of 10 limes (preferably key limes)

½ cup (120 ml) **heavy cream**

½ cup (120 ml) milk

½ vanilla bean, seeds scraped, pod reserved for another use

2 tablespoons (25 g) sugar

¼ cup (20 g) **unsweetened cocoa powder**

4 ounces (114 g) **semisweet chocolate**, chopped

½ cup (120 ml) **sweetened condensed milk**

3 tablespoons (45 ml) **lime juice**

2 tablespoons (7 g) graham cracker crumbs (optional)

Over medium heat, bring water and lime zest to a boil. Remove from heat, cover, and allow to infuse for 10 minutes. Return to medium heat and add cream, milk, vanilla seeds, and sugar. Bring to a simmer, then whisk in cocoa powder.

Remove from heat and add chopped chocolate. Let sit for 1 minute and then whisk to incorporate, using a rubber spatula to scrape any remaining chocolate from the bottom and sides of the saucepan.

Using an immersion blender, blend on medium speed for 1 minute. Add sweetened condensed milk and lime juice and blend for an additional 1 minute. Transfer to cups or mugs and sprinkle with graham cracker crumbs (if using).

Yield: 4 servings

If you don't have a micrograter, get one. It is the best tool to finely grate zest from a citrus fruit. It also works well for hard cheeses, ginger, garlic, and chocolate.

"PUMPKIN, PERFECTED"

PUMPKIN + FOIE GRAS + TRUFFLE

The sweet, luscious flesh of cooked pumpkin has a natural affinity for the luxury ingredients foie gras and truffle. The silken, sublime taste of foie gras meets the wild, earthy fragrance of truffle, wonderfully wrapped up with the pumpkin.

The application recipe takes this combination over the top, with a whole foie gras baked inside a pumpkin served with truffled toast. Other options for the combination include use as ravioli filling or as a creamy pumpkin soup with a chunk of Torchon of Foie Gras (page 254) and shaved truffles. These flavors work well with other seasonal fruits such as apples and pears, mushrooms, and cream.

PUMPKIN

For the application recipe, you'll need a three- to four-pound (1.4 to 1.8 kg) pumpkin, big enough to accommodate the whole foie gras, plus the liquid. Also, save the seeds to toast for a crunchy garnish—if you're feeling adventurous, sprinkle with a bit of truffle oil before baking. Look for pumpkin seed oil, which, with its intensely nutty taste, makes a great addition to salad dressings or drizzled over finished dishes.

FOIE GRAS

Foie gras' intensely rich livers come from force-feeding ducks or geese. The process—which lasts about 30 seconds two or three times a day, without any apparent suffering by the birds—is not without controversy. These points are debatable, so let your conscience guide you in deciding whether to consume it. Most foie gras available in the United States comes from ducks; goose foie gras is more common in France. You will generally have to buy a whole lobe, perfect for the application recipe.

TRUFFLE

Because of truffles' pungent flavor—white truffles have a stronger flavor than black—and high cost, cooks use them sparingly in food. They are available seasonally fresh or jarred in a light brine. Truffle-flavored oils and butters also are readily available and are a less expensive way to enjoy the intensity of truffles, especially because most dishes call for only a small amount.

"PUMPKIN, PERFECTED" RECIPE

Foie Gras–Stuffed Pumpkin with Truffled Toast

This is an impressive appetizer for any party; prepare it the day before serving.

Stuffed Pumpkin

1 **pumpkin** (3 to 4 pounds
   [1.4 to 1.8 kg]), top, seeds, and
   membranes removed
1 apple, peeled, cored, and chopped
¼ cup (60 ml) apple brandy
   (such as Calvados), divided
½ cup (120 ml) dry sherry
½ cup (120 ml) apple cider
Salt and pepper
1 whole **foie gras**, cleaned and deveined,
at room temperature

To Serve

1 baguette, cut into thin slices on
   the diagonal
Truffle oil for drizzling
Salt and pepper
1 apple, cut in quarters, cored, and
   thinly sliced

🎃 **This is a dish for fall,
when pumpkins are in
season. Try it with a warm
beverage.**

To prepare the stuffed pumpkin: Preheat oven to 275°F (135°C, or gas mark 1). Spread a large piece of foil on a sheet pan and place pumpkin on top.

Place the chopped apple in the bowl of a food processor, add 2 tablespoons (30 ml) apple brandy, and purée until smooth. Add the apple purée into the pumpkin with the sherry, apple cider, and remaining apple brandy. Sprinkle liquid with salt and pepper and then season both sides of the foie gras with salt and pepper. Add the foie gras to the pumpkin, with the large lobe on the bottom. Cover with the pumpkin top and wrap tightly with foil.

Bake for 2 hours until the internal temperature of the foie gras reaches 120°F (49°C). Remove the foil and allow the pumpkin to come room temperature. Refrigerate overnight.

To serve: Preheat oven to 350°F (180°C, or gas mark 4). Spread out the sliced baguette in a single layer on a sheet pan. Drizzle with truffle oil and season with salt and pepper. Bake until lightly golden, 4 to 5 minutes.

Remove pumpkin from the refrigerator 30 minutes before serving. Serve pumpkin with a hot spoon (place a cup of hot water nearby) for scooping. Serve with apple slices and a plate of truffled crostini.

Yield: 10 appetizer servings

"THE BASICS REDEFINED"

SALT + PEPPER + SUGAR

These three ingredients are a staple in every pantry and work together in countless recipes. Here, they act as a cure for the very rich meat of foie gras. Using these for this purpose both seasons and tenderizes the ingredient being cured. Items most often cured that work well with this combination include fish (salmon, anchovy, or snapper) and meats (beef, pork, and sausages). For a cured-fish variation on this combination, see Tea-and-Lime Cured Salmon (page 185) and substitute two tablespoons (12 g) of fresh ground pepper for the citrus zest.

SALT

The preferred salt for curing is coarse salt, which adheres to surfaces well and dissolves quickly. A brand such as Diamond Crystal is a good all-purpose salt for everyday cooking as well. Avoid curing salts, which have sugar and sodium nitrates mixed in. Instead, control your own seasoning and create your own blends.

PEPPER

To get the most impact from peppercorns, grind your own in small batches. Preground black pepper shaken from a container just doesn't provide the piquancy that fresh ground pepper delivers to this combination. If curing light-colored fish, use white peppercorns for a balanced appearance.

SUGAR

A standard granulated sugar is the most versatile choice for a cure blend. For curing hams, look to brown sugar or a turbinado sugar (such as Sugar in the Raw), which is less processed than white sugar and has a more complex flavor (though it does not have molasses added as brown sugar does).

"THE BASICS REDEFINED" RECIPE

Torchon of Foie Gras

This foie gras preparation produces an indescribably smooth, rich texture you must taste to believe. It is much easier to serve a large group of people when you slice foie gras instead of searing it over a hot stove (splattering fat all over your new outfit). Also, frozen foie gras turns out smoother with this technique than with searing. Come on ... go for it!

Special equipment required

1 whole foie gras

¼ cup (50 g) **sugar**

¼ cup (75 g) **coarse salt**
   (such as kosher)

2 tablespoons (12 g) fresh ground
   black pepper

2 cups (475 ml) chicken broth

1 cup (235 ml) sweet white wine
   (such as Muscat or Sauternes)

2 bay leaves

1 large sprig thyme

The richness of this foie gras goes perfectly with a sweet-tangy element. Try it with Rhubarb Compote (page 73) or Mango Chutney (page 15). A buttery brioche toast or crackers complete this awesome combination.

Separate the foie gras into two lobes and cut lobes across in 2-inch (5 cm) pieces. Place in a baking pan. Mix together sugar, salt, and pepper. Sprinkle over both sides of the foie gras, and press a piece of plastic wrap directly onto the surface of the foie gras to fully apply the seasoning. Refrigerate and allow to cure for 24 hours.

Prepare poaching broth by combining remaining ingredients in a large saucepan. Bring broth to a boil, then reduce to low. Add pieces of foie gras and cook for 2 minutes, or until they show signs of pliability (grab a piece and squeeze—remove as it starts to get soft). Pull out foie gras, put on a cold sheet pan, and put in the freezer for 5 minutes.

Press pieces of foie gras through a large, flat fine-mesh strainer onto a sheet pan (this removes any veins). Put the tray of foie gras in the refrigerator for 5 minutes.

Lay out two pieces of 12 x 12-inch (30 x 30 cm) cheesecloth on a work surface. Evenly divide the foie gras between the two pieces of cheesecloth into a log shape 8 inches (20 cm) long. Wrap the cheesecloth around it to form a tight cylinder, twisting the ends as you work. Roll the log back and forth to ensure no air pockets. Place on a sheet pan and refrigerate until solid. After 12 hours, wrap plastic around cheesecloth to slow down oxidation.

Torchons will keep for up to 2 days in the refrigerator. To serve, slice with a hot, wet knife. Season with additional salt and pepper or other spices as desired.

Yield: Two 8-inch (20 cm) logs

"CHOCOLATE, SQUARED"

CHOCOLATE + CHOCOLATE + CREAM

For the chocolate lovers of the world, the perfect accompaniment to chocolate is, well … chocolate. This is an unapologetically rich, intense chocolate pairing. Adding a cream element gilds the lily, taking this combination to the heights of "chocolate coma." The application recipe prepares serious chocolate soufflé with chocolate sauce and whipped cream. For another baking idea, use cocoa powder and chocolate chunks for an intense chocolate pound cake with crème anglaise.

CHOCOLATE

Chocolate ranges from unsweetened and semisweet to milk chocolate, with varying amounts of sugar versus cocoa creating each and affecting the flavor of the finished dish. Create interest in this combination by using different types of chocolate together—such as side-by-side mousses of white, dark, and milk chocolate.

CREAM

For this combination, whip cream with a little sugar for a whipped cream accent, set with gelatin for a smooth panna cotta, or make into a custard for a crème anglaise (to which you can also add chocolate) or a rich ice cream.

"CHOCOLATE, SQUARED" RECIPE

Chocolate Soufflé with Kahlua-Chocolate Sauce and Whipped Cream

This soufflé recipe is the ultimate chocolate dessert. First of all, it is something you can do completely ahead of time, with the exception of baking it. It also is utterly indestructible. This recipe may become your new favorite for entertaining.

Special equipment required

Soufflés

5 tablespoons (70 g) unsalted butter, softened, divided

1/2 cup (100 g), plus 1/3 cup (67 g) sugar, divided

8 ounces (228 g) **semisweet chocolate**, chopped

1 tablespoon (8 g) all-purpose flour

1/3 cup (80 ml) whole or 2 percent milk

3 eggs, separated, plus 1 additional egg white, divided

1 tablespoon (15 ml) vanilla extract

1/8 teaspoon cream of tartar

1/4 cup (25 g) powdered sugar, for dusting

Chocolate Sauce

10 ounces (285 g) **semisweet chocolate**, chopped

3/4 cup (175 m) **heavy cream**

2 tablespoons (30 ml) Kahlua (optional)

1 teaspoon (5 ml) vanilla extract

Salt

Whipped Cream

2 cups (475 ml) **heavy cream**

1/2 cup (100 g) sugar

1 teaspoon (5 ml) vanilla extract

Preheat oven to 375°F (190°C, or gas mark 5).

To prepare the soufflés: First, butter the bottom and sides of four 6-ounce (175 ml) soufflé ramekins using approximately 1/4 cup (55 g) butter. When buttering the sides, use your index and middle fingers to pull the butter straight up the sides. Be sure to butter the top edge of the ramekin as well.

Then coat with the granulated sugar. To do so, fill first ramekin with approximately 1/2 cup (100 g) sugar and tilt and rotate over a second ramekin, allowing the sugar to drop into it as it coats all sides. Pour any remaining sugar from the first ramekin into the second and repeat the process with the remaining ramekins. After sugaring final ramekin, pour remaining sugar into a shallow bowl. Dip the top edge of each ramekin into the sugar to coat.

Once you've prepared the ramekins, melt the chocolate in a large bowl set on top of a medium-size saucepan of simmering water. Do not allow the bowl to touch the water. Once the chocolate melts and smoothens, remove it from the heat and place it in a warm area, on top of your warm oven, if possible.

In the saucepan used to melt the chocolate, discard the hot water and wipe out the pan with a dry paper towel. Use this saucepan to melt 1 tablespoon (14 g) butter on medium heat. Add the flour and cook, stirring constantly with a rubber spatula, for 2 minutes.

Gradually add the milk, using a whisk until smooth. Keep whisking while occasionally using the rubber spatula to scrape the bottom and sides of the saucepan. Once the mixture thickens, remove it from heat and whisk in the egg yolks and vanilla. Scrape this mixture over the melted chocolate and fold until blended.

In a large bowl, beat egg whites with cream of tartar until foamy. Slowly and gradually add ⅓ cup (67 g) granulated sugar, and continue beating until stiff peaks form when you lift the beaters.

Fold the egg whites into the chocolate mixture. Fill your sugared ramekins with soufflé batter up to ¼ inch (²/₃ cm) from the top. Either cover with plastic wrap and refrigerate immediately for future use (up to 2 days ahead of serving) or put them on the sheet pan for baking.

To prepare chocolate sauce: Melt the chocolate with cream in a large bowl set on top of a medium-size saucepan of simmering water. Do not allow the bowl to touch the water. When the chocolate melts and smoothens, remove it from heat and add Kahlua (if using), vanilla, and a pinch of salt. If using immediately, put into small pitcher to pour onto soufflés. Sauce may be made up to 2 days ahead of serving and refrigerated. Warm sauce in the microwave before using.

To prepare whipped cream: Combine cream, sugar, and vanilla. Whip to medium firm peaks by hand or with a standing mixer. Whipped cream will stay for up to 1 hour covered with plastic wrap in the refrigerator.

To serve: Bake soufflés at 375°F (190°C, or gas mark 5) for 15 minutes. Soufflés are done once they rise high and have a crisp top. Dust with powdered sugar and use tongs to place hot ramekins in a small bowl or on a napkin-lined plate. Use a chopstick to poke 2 holes in top of each soufflé and fill with warm chocolate sauce. Scoop whipped cream on top and serve immediately.

Yield: 4 servings

Serving a soufflé is very dramatic and also has the impression of being temperamental. This recipe reliably rises and is very sturdy. Don't fear the soufflé! To introduce another flavor element and incorporate seasonal ingredients, try topping this dish with berries.

"AN UNRIVALED TRIO"

MARROW + CITRUS + PARSLEY

Go for it with this combination! No other eating experience compares to the rich flavor and unctuous texture that bone marrow offers. It's decadent and intense, so citrus lifts the flavor, and the mild freshness of parsley creates a welcome balance.

The application recipe features a paste of cold marrow to use atop a warm steak. For an alternative, roast the bones and spread cooked marrow on crostini or cook into a savory custard. This combination tastes especially great with beef or veal.

MARROW

The best marrow comes from the femur bone. Find what you need in Asian markets or at your local butcher. As a rule of thumb, expect each pound of bones to yield about 1 ounce or 2 tablespoons (30 g) of marrow. If you use the marrow of uncooked bones, use the remaining bones for stock. If you roast the bones first to get to the marrow, they're not really good for much after that—unless you have a dog, in which case the bones ensure your best-friend status.

CITRUS

Lemon is the classic choice for this combination. Use freshly squeezed juice to toss with parsley and shallot and serve with roasted marrow or use finely grated zest to season cold marrow. Orange makes a nice flavor variation. To increase the complexity of your dish's flavor profile, consider using lemon and orange together.

PARSLEY

Two parsley varieties dominate the market: curly or Italian flat-leaf. Choose the flat-leaf variety. It is more fragrant and less bitter than the curly variety. Also, there's no room for dried parsley in this combination—it can't provide the fresh herbal flavor needed to go with the rich marrow.

"AN UNRIVALED TRIO" RECIPE

Marrow Jam

Ask your butcher to split the bones lengthwise. Then scoop out the marrow and keep it covered in water in the refrigerator. This paste can intensify the pleasure of many dishes.

4 cloves garlic

1 tablespoon (15 ml) olive oil

Salt

1 shallot, diced

1 teaspoon (2 g) **lemon zest**

1 teaspoon (2 g) **orange zest**

2 tablespoons (8 g) chopped **parsley leaves**

1 teaspoon (1 g) fresh thyme leaves

2 tablespoons (1 ounce, or 30 g) cold **marrow**, chopped

Preheat oven to 350°F (180°C, or gas mark 4). Drizzle garlic with olive oil, season with salt, and roast in the oven for 20 minutes until soft. Let cool before proceeding.

Mash roasted garlic with a fork and mix with the remaining ingredients except for the marrow. Scatter the marrow on top of the mix and lightly combine. Cover and keep refrigerated until ready to use. Serve immediately.

Yield: ¼ **cup (80 g)**

Use this jam on steak hot off the grill or stir it into risotto to add a wonderful richness.

"RICH AND PROUD"

MUSHROOM + LIVER + CREAM

Let's just put this out there—this is a rich combination, and it is supposed to be. The mushrooms' meaty texture and full flavor are a natural with the actual meaty intensity of liver, and surrounding it all with cream creates lushness.

The application recipe prepares a creamy mushroom soup topped with a liver mousse crostini. Also, use the combination to sauté chicken livers with chanterelles in a thyme-cream sauce, for savory custard, or for pâté. These flavors go great with chicken, bacon, and fresh herbs.

MUSHROOM

To get the most mushroom flavor, use a combination of fresh and dried (which you'll need to reconstitute prior to using or simmer in liquid as part of the recipe). For complex flavor, incorporate more than one variety—shiitake, hedgehog, or chanterelles for fresh, porcini or morels for dried—into your dish.

LIVER

Chicken livers are by far the most easily accessible and least expensive liver to purchase. They make a nice choice for this combination. Try to find duck livers if you can; they are harder to get, but make for a richer liver flavor. Foie gras is specialty duck or goose liver that results from force-feeding the bird. It is much larger and richer than the average liver. It can be used in the combination, but sear it to highlight its wonderful, luscious qualities.

CREAM

Use cream as the base for a sauce, soup, or custard. Heavy cream is best for its richness, and a little goes a long way toward intense flavor. Low-fat dairy products do not add the required depth, but you can substitute crème fraîche or sour cream for creaminess.

"RICH AND PROUD" RECIPE

Creamy Mushroom Soup with Liver Mousse Crostini

Easily serve this intense soup as a meal with a salad. If using as a first course, keep the portion small so it won't fill up your guests.

1 pound (455 g) **fresh mushrooms**, cleaned (preferably a mix of shiitake, hedgehog, and chanterelles)

3 tablespoons (42 g) unsalted butter, divided

1 yellow onion, diced

2 stalks celery, diced

2 tablespoons (15 g) all-purpose flour

¼ cup (60 ml) dry white wine (such as Chardonnay)

3 cups (705 ml) chicken broth

2 bay leaves

1 sprig fresh thyme

1 ounce (28 g) **dried mushrooms**, crumbled (preferably porcini)

1 cup (235 ml) **heavy cream**

Salt and pepper

2 tablespoons (30 ml) cream sherry

2 tablespoons (8 g) finely chopped fresh flat-leaf parsley leaves

Chicken Liver Mousse (page 239—from Chicken Liver Mousse with Pea Purée and Sage Cream)

Baguette, thinly sliced into 8 crostini, drizzled with olive oil and toasted

Thinly slice 4 ounces (113 g) of the mushrooms and set aside. Chop the remaining mushrooms and keep separate from the sliced ones.

In a large stockpot, melt 1 tablespoon (14 g) butter over medium heat. Sauté the sliced mushrooms for 2 minutes until tender. Remove from the pan and set aside. Add the remaining butter with the remaining chopped mushrooms, onion, and celery. Cook for 1 minute and then add flour. Stir and cook for 5 minutes, until the onions and celery soften.

Deglaze the pan with the wine and then add the chicken broth. Cook for 3 minutes and then remove from heat. Allow to cool before adding to a blender. Blend until very smooth, then return to the pot. Add bay leaves, thyme, dried mushrooms, and cream. Bring to a simmer and cook for 10 minutes. Season with salt and pepper, then add sherry, reserved cooked mushrooms, and parsley. Simmer for 5 minutes on medium heat. Remove bay leaves.

Scoop chicken liver mousse on crostini and serve 2 crostini per bowl of soup, either on top (they will sink) or on the side. Serve immediately.

Yield: 4 servings

Richness like this is particularly satisfying as the temperature drops. This soup is great for a dinner party in the fall or winter months.

"A LITTLE BIT(E) OF PARADISE"

BANANA + CHOCOLATE + HAZELNUT

These three flavors together bring about a mouthwatering food lust that's hard to control. Don't even bother trying—let loose and bring them together. Banana's smooth sweetness plays nicely with rich mouth-coating chocolate and crunchy hazelnut.

These ingredients can work in many different ways: in banana cupcakes with chocolate-hazelnut frosting, a silky chocolate-banana pudding with chopped hazelnuts, or a bread pudding. They also work well with berries, dark rum, and any utensil you can get your hands on.

BANANA

Bananas are very versatile with this combination: straight out of the peel and sliced, puréed for puddings or breads, or sautéed to increase sweetness. Choose the application recipe and brûlée it to create a crunchy, sweet top layer.

CHOCOLATE

Though white chocolate is technically not chocolate because it contains no chocolate liqueur, it is nevertheless a nice choice for this combination, especially if you desire a pronounced sweet flavor. Milk chocolate, also quite sweet, makes for a richer, deep sweetness. Semisweet or bittersweet chocolate has less sugar and a more intense chocolate flavor.

HAZELNUT

Toast them to deepen their flavor—chopping and sprinkling them over any variation of this combination provides a satisfying bite. If your hazelnuts still have their skin, roast them in the oven and then use a towel to rub off the skins. Unless you are using the hazelnuts for a delicate baking application, don't worry if some skin remains. A little skin adds contrasting color and flavor.

"A LITTLE BIT(E) OF PARADISE" RECIPE

Nutella and Banana French Toast

Don't you love that certain breakfast favorites are really desserts in disguise? Cinnamon rolls, chocolate crois-sants, and now... French toast. This "sandwich" plays host to our combination ingredients, including a smear of Nutella (hazelnut spread) for good measure.

4 eggs

1½ cups (355 ml) milk

5 tablespoons (63 g) sugar, divided

¼ teaspoon ground cinnamon

½ teaspoon vanilla extract

8 slices bread, preferably brioche, cut
  1-inch (2.5 cm) thick

2 tablespoons (28 g) butter

2 **bananas**, peeled and sliced into
  thin rounds

½ cup (148 g) **Nutella**

¼ cup (29 g) toasted and chopped
  hazelnuts

You won't use the whole jar of Nutella for this recipe; not that it is hard to figure out how to use it up, but consider adding it to chocolate and cream for homemade truffles. Roll them in chopped hazelnuts for a consistent, crunchy flavor.

In a large bowl, whisk together eggs, milk, 1 tablespoon sugar (13 g), cinnamon, and vanilla extract. Heat a large skillet over medium-high heat.

Dip each piece of bread into egg mixture, turning to coat both sides and allowing mixture to penetrate the bread. Add butter to skillet and cook bread until browned, 2 to 3 minutes and then turn and brown the other side. Remove and place on a sheet pan.

Preheat oven broiler. Place sliced bananas on 4 pieces of the bread you just made into French toast. Sprinkle evenly with remaining sugar and place under the broiler until browned and caramelized. (Alternatively, use a handheld brûlée torch if you have one.) Let cool for 5 minutes.

Spread Nutella on remaining 4 slices of bread and sprinkle with chopped hazelnuts. Make a sandwich by placing Nutella bread on top of the banana-covered bread. Slice on the diagonal to form 4 small triangular pieces. Serve immediately.

Yield: 4 servings

"PEAR-ING UP THE CLASSICS"

PEAR + CHOCOLATE + HAZELNUT

The combination of chocolate and nut appears in countless classic candy bars. And Nutella has brought hazelnut with chocolate to the masses. Placing pear with this decadent mouthful cuts through the richness, adding a bright note and a pleasing texture contrast. Melting the chocolate and cooking the pear softens the bite, adding crunch to the already-crunchy hazelnut. For a different application, make your own chocolate bar by using high-quality bittersweet chocolate tempered with pear liqueur and hazelnuts.

PEAR

Ripeness is critical when eating pears raw. This fruit poaches nicely, but handle with care any cooking methods you try on pears because of their high water content. Anjou pears poach well, and this is one instance when buying slightly underripe fruit will benefit the dish; it won't fall apart while cooking. For a different effect (and a sweet bite), dip dried pears into chocolate and sprinkle them with crushed hazelnuts.

CHOCOLATE

Chocolate ranges from unsweetened and semisweet to milk chocolate, with varying amounts of sugar versus cocoa creating each and affecting the flavor of the finished dish. Use of a bittersweet chocolate (less sweet) results in a more refined, adult taste. Milk chocolate has much more sugar, which means it satisfies the sweetest of teeth.

HAZELNUT

Toast and roast hazelnuts to deepen their flavor. In this combination, use them as a textural contrast by sprinkling them atop pears. Other options: Grind them to and combine with chocolate for a rich sauce or use as a hazelnut flour for a baked chocolate bundt cake with pear compote.

"PEAR-ING UP THE CLASSICS" RECIPE

Poached Pears with Chocolate Sauce and Hazelnuts

Feel free to experiment with your seasoning in the poaching liquid—cinnamon, nutmeg, or even curry can add an interesting underlying flavor component.

1 cup (235 ml) water

1 cup (235 ml) sweet wine (such as Riesling or Muscat)

½ cup (100 g) sugar

½ vanilla bean, sliced lengthwise and seeds scraped

Juice of 1 lemon

2 slightly underripe **pears**, peeled, halved, and cored

2 ounces (57 g) bittersweet **chocolate**, chopped

½ cup (68 g) **roasted hazelnuts**, chopped

Vanilla ice cream (optional)

Combine first five ingredients (through the lemon juice) in a medium-size saucepan and bring to a boil. Add pears, reduce heat to a simmer, and cover. Cook for 10 minutes until pears are tender and translucent. Remove pears from poaching liquid and allow to drain on a paper towel-lined plate. You may prepare this 1 day ahead; cover with plastic wrap and refrigerate.

Bring poaching liquid back to a boil and cook for 10 to 15 minutes to reduce the liquid to about ¾ cup (175 ml). Add chopped chocolate to saucepan, swirl to cover, and allow to sit for 1 minute. Whisk until smooth.

Slice pears thinly and fan out on a serving plate. Drizzle with chocolate sauce and sprinkle with hazelnuts. Serve immediately with a side of vanilla ice cream (if using).

Yield: 4 servings

Thankfully, chocolate is always in season. Pears, however, peak during the fall, so take advantage of the season's bounty. This combination is almost guaranteed to please most palates, so it is perfect for a large, diverse group.

"WHOLESOME INNOCENCE"

LEMON + RASPBERRY + EGG

Tangy fresh lemon is an exciting palate teaser that interacts nicely with sweet raspberries. The two flavors, which tango to an unknown beat, have a purpose when eggs enter. The egg yolk transforms the ingredients into a tangy-sweet curd, and the egg white creates a gorgeous baked meringue for the application recipe. Plus, the egg offers richness to the flavors and enables many other baked applications. Try muffins, pies, or mousses for alternatives.

LEMON

For this combination, which frequently produces baked items and desserts, you use the juice and zest most frequently. A citrus juicer makes for fast juicing, and a micrograter removes the zest so that it is finely grated and easy to use. To get the most liquid out of your lemon, roll it on the counter with pressure before you juice and look for lemons that feel heavy for their size.

RASPBERRY

Cooking reduces the raspberry's flavor, so use fresh or barely warmed for the most impact. When buying fresh at the grocery store, look at the pad lining the bottom of the plastic container—it should be unstained for maximum freshness. If using raspberries for a sauce application only, consider frozen berries, which often have a more consistent flavor than fresh berries shipped long distances. For a smooth sauce or sorbet, strain the mixture to remove the seeds.

EGG

For this combination, effectively use yolks as thickeners (as in the curd in the application recipe) and soufflé bases, whites for meringues and soufflés, and whole eggs for quick breads and cakes. Local or free-range eggs often have yolks with much deeper yellow color and fuller flavor than the average egg. They are worth pursuing for your cooking, if possible.

Lemon-Raspberry Meringue Nests

This dish is based on the elegant Pavlova dessert—swirling baked meringues topped with lemon-raspberry curd, whipped cream, and fresh fruit. This is a sweet dessert and makes for an impressive presentation at the end of any meal.

Meringue Nests

3 **egg whites** (yolks reserved for the curd below), at room temperature

Pinch cream of tartar

1¼ cups (250 g) sugar, divided

1 teaspoon (5 ml) vanilla extract

To make the meringues: Preheat oven to 200°F (93°C, or gas mark ¼). On a piece of parchment paper, use a pencil to trace six 4-inch (10 cm) circles, at least 2 inches (5 cm) apart. This makes 2 extra nests to account for potential breakage—you can crush ones you may not use to sprinkle on top of the finished dessert. Turn over parchment paper and place on a sheet pan.

In the bowl of a standing mixer, beat the egg whites with the cream of tartar until soft peaks form. With the motor running, add ¼ cup (50 g) sugar in a steady stream. Continue to beat on high until the whites become very stiff and glossy and then beat in the vanilla extract.

Using the circles as a guide, gently shape the whites into a nest shape with the back of a large spoon. Smooth the top, creating a slight indention in the center. Bake until the meringues crisp on the outside and detach easily from the parchment, 1 to 1½ hours. Turn off the oven and allow to cool completely in the oven. Peel paper off the bottoms and use immediately or store tightly wrapped.

Lemon Curd

Zest from 1 **lemon**

8 **egg yolks**

¼ cup (60 ml) fresh **lemon juice**

Salt

10 tablespoons (140 g) unsalted butter, cut into 10 pieces

1 cup (125 g) raspberries

To Serve

1 cup (125 g) **raspberries**

1 cup (235 ml) heavy cream, whipped to soft peaks

To prepare the curd: In a saucepan off heat, whisk together the remaining 1 cup (200 g) sugar, zest, and egg yolks. Whisk in the lemon juice and a large pinch of salt, using a rubber spatula to scrape the edges.

Add the butter and place pan over medium-high heat. Whisk constantly until the butter melts and the mixture thickens but is not boiling. Pass through a fine mesh strainer, press a piece of plastic wrap directly on the surface, and refrigerate until cold. Purée 1 cup (125 g) raspberries in a blender and push through a fine mesh strainer to remove the seeds. Fold raspberry purée in chilled curd prior to assembly.

To assemble the nests: Top nests with lemon-raspberry curd, a dollop of whipped cream (from the heavy cream), and a scattering of the remaining fresh raspberries. Serve immediately.

Yield: 4 servings

Make your curd and nests the day before serving. However, know that high humidity can adversely affect the nests, so store them airtight in zippered plastic bags or containers with tight lids. Feel free to add the fruit of your choice to the raspberries—kiwi makes a nice color contrast and captures the dessert's Australian origins.

"MELODY OF CONTRASTS"

CHOCOLATE + CHEESE + BERRY

Chocolate and cheese both taste rich but have very different flavors. They come together for many dessert applications and get a nice fresh fruit contrast when paired with berries. The most room for variation in this combination comes with cheese selection. The application recipe combines goat cheese with chocolate for a rich fondue with strawberries. Other options: Use ricotta or cream cheese with cocoa powder for a cheesecake topped with raspberries or scoop out the inside of a strawberry with a melon baller and fill with mascarpone-chocolate ganache. These flavors taste great with nuts and honey.

CHOCOLATE

A bittersweet chocolate provides the most intense chocolate flavor. A white chocolate contrasts beautifully with dark chocolate or, if used alone, offers a sweeter effect. The application recipe directs you to melt the chocolate in a saucepan on the stove, but chocolate also melts nicely in the microwave. Be sure to set the microwave for short intervals (30 seconds to 1 minute) and stir frequently so the chocolate doesn't burn.

CHEESE

With the chocolate in the mix, this combination leans toward dessert applications. Cheese options we suggest include mascarpone, fresh goat cheese, cream cheese, or soft-ripened cheeses such as Brie. Also, for a different twist on sweet canapés, look for a chocolate goat cheese made by the American company Westfield Farm.

BERRY

Ripeness is critical for getting the sweetest strawberry possible. Buy ripe strawberries to use within a day or two of purchase and leave them at room temperature. For purées, push thorough a fine mesh strainer to remove those pesky seeds from your finished dish. It's a small step but one that results in a more refined preparation.

"MELODY OF CONTRASTS" RECIPE

Heavenly Chocolate Fondue with Strawberries

Fresh goat cheese is melted in this dessert fondue, producing an intensely chocolate flavor (not cheesy)—a perfect dip for fresh strawberries.

24 ounces (684 g) **bittersweet chocolate**
   (preferably at least 58 percent cocoa solids), chopped
½ cup (120 ml) heavy cream
2 tablespoons (30 g) crème fraîche (or sour cream)
8 ounces (227 g) **fresh goat cheese**
1 teaspoon (5 ml) vanilla extract
4 cups (580 g) **strawberries**, hulled

Combine the chocolate, cream, crème fraîche, and goat cheese in a saucepan. Melt over low heat, stirring constantly with a rubber spatula until the fondue melts and smoothens. Stir in the vanilla extract. Transfer to a fondue pot or other small pot and serve with fresh strawberries.

Yield: 2 cups (500 g), plus 4 cups (580 g) strawberries

For variety, add other berries or fruit to the strawberries—bananas, grapes, pineapple. etc. Or try Savory Shortbread (page 39) for further flavor intensity.

"A RICH, TICKLISH DELIGHT"

EGG + LIQUOR + NUTMEG

Eggs' richness and versatility make them a unique choice for beverages. Generally, eggs in drinks aren't cooked, so remember the blanket warnings against consuming raw egg products. That said, a burst of liquor mingling with the fullness of egg and intense pungent fragrance of nutmeg produces a delicious mouthful.

The application recipe prepares a classic eggnog, perfect for holiday indulging. For a nondrink application, use the combination in baked goods such as a spiced rum banana bread or bourbon-laced fruitcake. These flavors go great with pumpkin, banana, chocolate, and cinnamon.

EGG

Egg whites often get mixed into cocktails such as whiskey sours and pisco sours as a thickening agent, creating a fluffy, foamy texture and mouthfeel. If you are squeamish about using fresh egg whites and the small chance of a foodborne illness, use powdered egg whites (not meringue powder). They are not a synthetic product, but rather made from egg whites whose water has been removed. Fresh whites are always the best, however, so take care to find local, organic eggs for raw applications.

LIQUOR

The depth of this application can handle a full-flavored liquor. Try whiskey, bourbon, Cognac, or dark/spiced rum. Nut liqueur, such as the hazelnut-flavored Frangelico, is also a nice match. Don't be afraid to add a healthy amount of liquor to this combination—the richness of the egg and spice of nutmeg enjoy the boost of the booze.

NUTMEG

Use freshly ground nutmeg, ground with a micrograter or a specially made nutmeg grinder, similar to that used for peppercorn. If you buy preground nutmeg, buy it in small amounts, as the flavor fades over time.

"A RICH, TICKLISH DELIGHT" RECIPE

Liquor-Laced, Old-School Eggnog

If you have only tasted eggnog from a carton, you may not recognize this rich and creamy version.

4 **eggs**, yolks and whites separated

1 cup packed (225 g) brown sugar

½ cup (120 ml) **bourbon**

½ cup (120 ml) **Cognac**

½ cup (120 ml) **dark rum**

Salt

½ teaspoon **freshly grated nutmeg**, plus more for garnish

2 cups (475 ml) milk (preferably whole)

2 cups (475 ml) heavy cream

1 teaspoon (5 ml) vanilla extract

Whisk the egg yolks in a large bowl while slowly adding the brown sugar. Continue to whisk until the brown sugar dissolves (you can also use a standing mixer for this). Slowly pour in the bourbon, Cognac, and rum. Add a large pinch of salt, nutmeg, and milk and whisk until smooth. At this point (for a do-ahead option), you can cover and refrigerate overnight.

Beat the egg whites on high speed until soft peaks form. In a separate bowl, whip the heavy cream and vanilla extract to soft peaks. Fold the egg whites and whipped cream together; then fold into egg yolk mixture. Chill until ready to serve. Before serving, whisk to make the eggnog frothy. Ladle into cups and sprinkle with nutmeg.

Yield: 12 servings

For a variation, use chocolate milk and sprinkle with grated nutmeg and cocoa powder.

"A RAUCOUS GOOD TIME"

CHOCOLATE + PEANUT BUTTER + MALT

Why not combine three wonderfully rich flavors to produce an over-the-top combination? Each flavor can stand comfortably on its own, but when put together, they become dessert intensity—in the best way.

This chocolaty, peanuty, malty combination hits almost every potential sweet-hankering button. The application recipe prepares a thick, rich chocolate-peanut butter malt shake. Also use the combination for baked goods such as brownies, cookies, or pancakes.

CHOCOLATE

Chocolates range from unsweetened and semisweet to milk chocolate, with varying amounts of sugar versus cocoa varies creating each and affecting the flavor of the final dish. Use any variety in this combination; let the flavor of the finished dish guide you (milk for more sweetness, bittersweet for more chocolate intensity). White chocolate is also a good option for this combination; try substituting chunks of it for dark chocolate in cookies or brownies.

PEANUT BUTTER

Choose a smooth, natural peanut butter for cooking. You can always add roasted chopped nuts to create a "chunky" texture, but smooth is more versatile for cooking applications. We prefer natural, unsweetened peanut butter for its pure flavor. Avoid brands loaded with sugar and preservatives. Instead, go for one that's minimally processed.

MALT

Malted milk powder is the most readily available malt product in grocery stores (such as Carnation brand) and combines malted wheat or barley flour with powdered milk for use in milk and ice cream. For a pure malt flavor for baking applications, seek out pure barley malt powder or syrup (which has no powdered milk or other fillers) in health food stores.

"A RAUCOUS GOOD TIME" RECIPE

Triple Threat Milkshake

This rich milkshake blends a chocolate-peanut butter ice cream with a rich malt milk. Sprinkle with chopped peanut butter cups and malted milk balls—if you dare.

Special equipment required

7 ounces (200 g) **milk chocolate,**
  chopped
4½ cups (1 L) heavy whipping cream
¾ cup (150 g) sugar
¼ cup (65 g) smooth **peanut butter**
1½ cups (355 ml) milk
⅓ cup (84 g) **malted milk powder**
Finely **chopped peanut butter cups**
  (optional)
Finely **chopped malted milk balls**
  (optional)

Place the chopped chocolate in a large bowl. Combine the cream and sugar in a saucepan and bring to a simmer, stirring to dissolve the sugar. Once it starts to boil, remove from heat and pour over the chocolate. Let stand for 1 minute and then whisk together until smooth.

Add the peanut butter and whisk until smooth. Place over an ice bath and stir occasionally until cold. Add to ice cream machine and process according to the manufacturer's instructions.

Add finished ice cream to a blender with the milk and malted milk powder. Blend until smooth, pour into tall glasses, and sprinkle with candy (if using).

Yield: 4 shakes

This can be a fun dessert to serve at a seated dinner party. Or pull it out when you have a family gathering. Or simply enjoy one when you have the house to yourself. The possibilities are endless.

CONCLUSION

Combining flavors resulting in dynamic and satisfying tastes should be a lifelong pursuit for any cook. This book is but a small taste of how one can think about bringing together ingredients for delicious, intensely flavored dishes.

Try new tastes, develop new skills, but above all, have fun and enjoy the process. Hopefully, along the way, you will expand your food world and crave to create more stimulating, mouthwatering combinations of your own.

RESOURCES

Specialty Equipment
J. B. Prince
800.473.0577
www.jbprince.com

Specialty Meats and Foie Gras
D'Artagnan
800.327.8246
800.DARTAGN
www.dartagnan.com

Mushrooms, Truffles, and Vinegars
Fresh & Wild
800.222.5578
www.freshwild.com

Chocolate Products
Scharffen Berger Chocolate Maker
800.930.4528
www.scharffenberger.com

ACKNOWLEDGMENTS

This book is the result of the efforts of many people. Our staff at Krause Dining during the time this book was written took up the slack in many ways—thank you Zach Hangauer, Janai Tate, Richard Garcia, and Nick Gilbert. Our business partners and friends Simon and Codi Bates provided support, friendship, and recipes. Our customers, going back to our Topeka days to Krause Dining and the Burger Stand, are a constant source of inspiration and community.

Thank you to my mom, Katie Krider, for her shrewd editorial skills and finding her inner poet for many of the titles. Shelly Gaudreau and Sandra Moran also provided important encouragement for my writing pursuits.

The people at Fair Winds Press were instrumental in shaping the vision for the finished book, particularly Jill Alexander. The creative vision and professionalism of Rosalind Loeb Wanke, Luciana Pamplone, and Alex Troesch was much appreciated in creating a beautiful book. But there would be no book without our agent Neil Salkind, whose amazing perseverance on our behalf and whose constant stream of positive words cannot be matched. Our daughters Lauren, Emma, and Cameron are sources of abundant joy in our lives. Thank you girls for enduring the extra "homework" we had for many months.